Mosaic of a Scientific Life

István Hargittai

Mosaic of a Scientific Life

István Hargittai
Department of Inorganic and Analytical Chemistry
Budapest University of Technology and Economics
Budapest, Hungary

ISBN 978-3-030-34768-0   ISBN 978-3-030-34766-6   (eBook)
https://doi.org/10.1007/978-3-030-34766-6

This edition in English is based on the book in Hungarian, István Hargittai, *Mozaikokból egy élet* (Budapest: Akadémiai Kiadó, 2019).

Cover illustration: Front cover and title page art by István Orosz, Budapest, 2019. He is an artist, graphic designer, and film director, known also for mathematically inspired works (see a chapter bearing his name).

This Springer imprint is published by the registered company Springer Nature Switzerland AG.
The registered company address is: Gewerbestrasse 11, 6330 Cham, Switzerland

"István Hargittai has led an extraordinary life in science, which has brought him into contact with some of the great scientific figures of the twentieth century. These vignettes, both of acknowledged giants and of some individuals who deserve to be better known, create a chronicle that it is a delight to see recorded for posterity—a record of how science reflects all the vicissitudes of that turbulent century, and of how diverse individuals have found their own paths through it. This is a warm personal journey as well as a valuable historical document."

—Philip Ball, *science writer (London), a former long-time editor of* Nature, *author of* Patterns in Nature *and* Beyond Weird

"This wonderful book is a tapestry as well as a mosaic. Hargittai skillfully weaves his own notable life as a scientist and a writer into the lives of an international Who's Who of other scientists, writers, and thinkers."

—Kenneth W. Ford, *author of* 101 Quantum Questions *and* Building the H Bomb

"István Hargittai is best known for the stories he tells of others, brilliant scientists, who shaped our understanding of the world. He has succeeded in doing that in a series of many interviews. Here we are treated to learning about the interviewer in a fascinating account of a life spent illuminating what is scientific creativity and how the pursuit of knowledge is really done."

—Richard N. Zare, *Wolf Prize and King Faisal Prize laureate; ACS Priestley Medalist, Stanford University*

"The author summarizes a broad panorama of European scientists and artists that he has met on various occasions. The spectacle is engrossing with much of the discussions specifically with Hungarian individuals. Several of the people involved, including the author, are survivors of the Holocaust and their determined and impressive achievements mark them out as leaders in any time and place."

—Sidney Altman, *Nobel laureate, Yale University*

"This is a different kind of biography. As Goethe said 'Tell me whom you consort with and I will tell you who you are'. István Hargittai has had a rich and at times turbulent life. Besides being a qualified physical chemist interested in symmetry he is an empirical real time historian. In this book many of his personal and professional friends have also been included. The book's driving force is in good Pythian tradition to learn to 'know thyself.' We do this by interpreting the reflections of ourselves in others."

—Erling Norrby, *former member of the Nobel Committee for Physiology or Medicine and permanent secretary of the Royal Swedish Academy of Sciences*

"This mosaic is a very vivid account of Hargittai's personal insight and impressions of great twentieth century scientific minds. I could not stop reading the almost 50 short stories told by the man who met them all for long discussions."

—Dan Shechtman, *Nobel laureate, Technion*

"A 'tour de force' is the thought that comes to mind when reading this latest Hargittai book. A monumental collection of anecdotes and memories of 20th century scientists told in the most insightful and optimistic way. He is a master of the biographical sketch."

—Professor Sir Salvador Moncada, *FRS, FMedSci, HonFBPhS, University of Manchester Cancer Research Centre*

"What good fortune for us that István Hargittai has had over the years the extraordinary talent of making friends of shapers of the spirit, artists and scientists, people who have added value to the world. In this book he brings many of them, a good number of Hungarians among them, close to us. In fascinating personal vignettes, Hargittai's sketches connect up our world."

—Roald Hoffmann, *Nobel laureate chemist and writer*

"István Hargittai has written a series of interesting accounts of interviews he has had with a wide range of academics and public figures. These accounts not only reveal much about a broad span of Twentieth Century intellectuals, but also about István Hargittai himself. He has written an unusual autobiography in which we learn about him through his meetings with other people."

—Sir Paul M. Nurse, *Nobel laureate, Chief Executive and Director of the Francis Crick Institute (London), and former President of the Royal Society*

"István Hargittai has written many highly engaging books with interviews and commentaries about science, scientists and the scientific enterprise. His latest contribution "Mosaic" follows in this excellent tradition. His descriptions, anecdotes and philosophical observations are partly autobiographical and partly journalistic, but always gripping, readable and original. I hope the readers will enjoy it as much as I have."

—Richard Henderson, *Nobel laureate, MRC Laboratory of Molecular Biology, Cambridge*

for Magdi

Foreword by Agnes Heller[1]

The mosaic of this book is based on meetings. The author, István Hargittai, met with scientists, Nobel laureates, and others, among them friends who were also mostly scientists, during the last decades of the twentieth century and the first of this century. The pattern of the mosaic is alphabetic rather than chronological. The "life" emerging from this mosaic is the author's life. The mosaic tells us who the author is and what he is, and what experiences made him into what he has become; the "who" and the "what" appear in inseparable unison.

Already in his childhood, loyalty, persistence, and curiosity characterized him and these traits have endured for his entire career in science. Thus, for example, first he had become curious in the occurrences of symmetry, and then he stayed with it, and sought connection to those scholars who shared his interest. His persistent dedication then produced books on this topic. His loyalty to his interest did not let him stop at the borders of his research areas; rather, he broadened his inquiries to embrace the great variety of everyday manifestations of symmetry. I was taken by what he had to say about the buckminsterfullerene molecule, and not because I understood exactly what it was, but because I learned that it has been considered to be the most symmetrical and most beautiful among the molecules. What should a philosopher think upon hearing such a designation? That it was almost commonplace for the ancient Greeks that symmetry was the source of beauty, which was first questioned by Plotinus.

Hargittai stresses the same three traits mentioned above, loyalty, persistence, and curiosity in his presentation of scientists—in this book of conversations or a gallery of portraits. For him, it must have been curiosity above all that led him along the path of getting to know so many famous scientists. What else could have lured a renowned scientist, university professor, and recognized author of professional monographs to interview others? What else could have induced him to spend the considerable time necessary for preparing himself, and not only in his own field but in areas that were far from his immediate interest and without expecting them to become his professional interest. This project often involved travels to faraway destinations. Why was he wasting his time? Obviously, he did not consider it a waste. His curiosity—both professional and human—was driving him: to learn about the individual through the accomplishment and about the accomplishment through the individual. This was no journalistic curiosity, because the curiosity of the scholar is not directed toward "anything"; rather, it has specific purpose. This curiosity is aiming at understanding the genesis, the process of becoming—in this case, of becoming a significant scientist.

There are as many roads to becoming a scientist as there are scientists—these paths are impossible to repeat, even for those who already had done it before. The "product," that is, the life of a scientist or a professor, focuses on research or teaching, or both. In addition, the process of the formation of the individual and the individual's role in the world, in history, and the accumulating experience are all also of importance. There must be something common among

[1] Agnes Heller (1929–2019), world-renowned Hungarian philosopher and fiercely independent thinker, Professor Emeritus at the New School for Social Research in New York, died on July 19, 2019—drowned in Lake Balaton while swimming at Balatonalmádi.

the individuals who belong to the same world and the same era. The interview, the conversation, is a way to uncover this commonality. The emerging knowledge is not a scientific discovery; rather, it is a discovery in historical anthropology.

Aristotle's *Metaphysics* begins with the following sentence: "All men by nature desire to know." The question is what and when, and how? How can the desired knowledge be found even if it is never absolute? István Hargittai is seeking the answer to these complex questions in his interviews and other conversations. Max Weber at the start of the twentieth century added that the modern individual should never consider any knowledge absolute, because the next generation will pose new questions from the previous answers. This process never ends, ever. This is what Popper expressed when he said that scientific truth is true only because it is falsifiable.

In spite of Weber's skeptical comment, almost all of Hargittai's stories are success stories. These scientists were not looking for "the truth"; they were looking for specific knowledge that was the subject of their research. If they did succeed—and especially if their success was crowned with a Nobel Prize—they could turn elsewhere. Having fulfilled their task, they could change the direction of their inquiry, engage in a hobby, get involved in pedagogy, author books or poetry, do gardening, or embark on world travel. Hargittai values such changes, and he followed such a path as did George Klein, László Bitó, and many others. This already justifies that not all the portraits in the book are scientists'. There are others with whom he has been friends, fellow students, or colleagues, or with whom he merely had conversations: from the writer György Dalos to the former president of Hungary, Árpád Göncz. The majority are, however, scientists.

With proper tact, to be sure, but Hargittai considers the traits of the individuals to whom he introduces the reader, and not only their success stories. He has preferences and he does not shy away from letting us know what they are. While recognizing greatness in science, he does not like vanity, jealousy, and careerism, adulation of the powerful, or seeking popularity. He places Crick higher in human values than Watson in the Watson–Crick duo although he equally values Watson's oeuvre and cannot identify with all of Crick's opinions outside the realm of his science. As for the political profiles of scientists, his stand is unambiguous concerning the services scientists perform for political causes, that is, in connection with the roles they play in defending a cause or display loyalty to one or another.

Hargittai considers his meetings with scientists and their science to be components of the mosaic of his life as they truly are. This is because science has been on center stage all his life. It could not be accidental that his wife, Magdolna, is also a scientist, and they have coauthored several books. It is no mere chance as nothing has been a mere chance in Hargittai's life and neither in the lives of his interlocutors. Chance has played a role in all their lives. A chance did not just stay a chance, because they did something with it. A chance is a good servant if its master knows how to utilize it.

The parallels between Hargittai's life experience and that of his partners play a distinguished role in this book. Hargittai lost his father when he was one-year-old: He was killed in slave labor. So when Hargittai meets someone who also lost a father at an early age, he discovers similarities. There are holocaust survivors among his partners and we learn a great deal about István's and his mother's "lucky" fate as they were deported to a *lager* (camp) in Austria. When he meets with Jews who had fled Hungary in time, but whose family members were murdered in Auschwitz, we learn about the experience that has impacted the rest of their lives. The activities of émigré scientists who played a determining role in the war efforts of the USA against Nazism, among them the Jewish refugee scientists, are given distinguished attention in the book. They include the five "Martians" (Theodore von Kármán, Leo Szilard, Eugene P. Wigner, John von Neumann, and Edward Teller) about whom Hargittai had written a monograph. From among the émigré scientists, Peter Lax and George A. Olah also figure in separate chapters. Some praise their excellent teachers and Hargittai also remembers his teachers in high school and university, especially those who supported and encouraged him at the critical times when his "class-alien" social status placed him in disadvantaged situations for his studies. These reminiscences remind me the bon mot of my late friend András Pernye. Every significant person had outstanding teachers, he said, and he meant it with irony. Significant people can learn a great deal from almost anybody.

On the basis of Hargittai's mosaic of his personal and scientific life, I could compose two further patterns. One would be the history of the twentieth century and the other the science history of the same time period. The two would not be the same even though the overlaps would be substantial.

Hargittai was a child and youth and became a scholar in Hungary—in a small town and in Budapest. As a persecuted child, he had to wear the yellow star, was condemned to death, and was deported. As a teenager he was stigmatized with the "class-alien" label and had to struggle for the possibility of studying. This did not crush him; on the contrary, he only became more persistent. Later, still under the Kádár regime, he received a scholarship to continue his university studies at the Lomonosov University in Moscow. Thus, the reader gets some introduction to Moscow and the Soviet Union under Khrushchev and then Brezhnev. He conveys his experience with outstanding Soviet scientists who did everything for protecting their independence and their science. He writes about some with whom he never met, such as the physicist and human rights activist Andrei Sakharov, and others, outside of science, such as the sculptor Ernst Neizvestny whom he met in person only decades later. He remarks about Neizvestny's grave memorial of Khrushchev's tomb that it is an excellent example of anti-symmetry. When he mentions Imre Kertész, we learn that *The Union Jack* is his favorite among Kertész's works. I add that even though this is not my favorite Kertész work, *The Union Jack* is the truest story ever written about the 1956 Revolution.

Then, the reader is introduced to the USA where the author spent so much time in so many places, doing research and teaching, and from whence the largest number of Nobel laureates "came," starting from the 1930s and on. We learn about the school system that, at the time, made it possible even for the poorest, but gifted, child to receive the best education at the highest level.

From these components of the mosaic, the reader can compose the science history of the last half century. The Nobel Prize, according to Hargittai, does not always reward the most outstanding achievements, and not all outstanding achievements are rewarded with a Nobel Prize—and he refers to the Bible, "many are called but few are chosen." This award plays a different role in the sciences than in literature. One writer is awarded the Nobel Prize in Literature annually and much pain is taken to distribute it properly among continents and languages. In science, there are three Nobel Prizes and up to a total of nine scientists may become recipients. Considerations, such as the mother tongue of the laureate, play no role. In contrast, the recipients of the literature prize are known much beyond a narrow circle of colleagues, whereas this is not the case for most of the science laureates. This was not always so. I have myself been familiar, to the extent a non-specialist might be, with Einstein's or Heisenberg's theories. However, I heard about more than half of the Nobel laureates figuring in this book, for the first time. I have become aware of them only because I read about them in this book, and I learned about their greatness from what they themselves revealed about their dedication, perseverance, loyalty, and insatiable curiosity. I know from one of Hargittai's scientist friends, George Klein, that there are many stupid people even among the Nobel laureates. It seems that Hargittai did not meet them or he just dreamed his own thoughts into them.

As I said, the reader can compose from these components of Hargittai's mosaic the science history of the last half century (or even 70 years). This is one of the three patterns (the author's life story, the history, and the science history of Europe and America) emerging here that have overlaps but cannot substitute one for another.

As all histories, this is also about the past.

It may well be the limitations of my vision, but my impression is that the science history composed from this mosaic is colored by nostalgia. It is as if the author would be saying farewell to the world of science about which he writes. My impression was strengthened by one of his remarks and his reference to one of Arthur Koestler's later books discussing science theory. This is Hargittai's remark: "We live in an era when instant results are expected and there is insufficient tolerance for a universal and inevitably skeptical approach to the big questions of science." This is a clear indication of the diminishing possibility for a paradigm change. In the book referred to by Hargittai, Koestler writes about the role of the unconscious in the great and innovative discoveries and that it often comes from "bisociation," that is, the connection between two independent phenomena linked only by the thoughts of the discoverer. Koestler supports his proposition with numerous examples and testimonials. What is the prerequisite for a great and innovative discovery to happen? For this, individual scientists need to attack the tasks they consider exciting individually: a team of researchers have no collective subconscious. Also needed is time; deadlines cannot produce discoveries. Needed also is a broader approach and an extended thought process to lead to results of greater significance rather than to small additions to the already existing knowledge. If I interpret it correctly, Hargittai's feeling of nostalgia originates from the fast disappearing conditions for the kind of research that might produce even paradigm changes. I must add that this state of affairs characterizes not only the "natural" sciences but all bureaucratizing sciences.

I have composed three patterns; some readers might be able to produce a fourth. Try it!

Budapest   Agnes Heller
Spring 2019

Preface

Looking back on our life, we can see it as a continuum in which every step follows from the previous one in an orderly manner. However, we can also see it as if it had been composed as a mosaic. There may be natural connections between the parts, but it may be that we only recognize the connections in hindsight. I will give a description in both ways.

It is quite noticeable that I have met many Nobel laureates during my career. This does not determine anything about me, yet a rather consistent narrative can be offered as an explanation. My favorite subject used to be mathematics. When I won a mathematical competition, I received a chemistry book as a prize, and this started my love affair with chemistry. As a chemist I recognized the importance and utility of symmetry and started publishing some papers in a periodical devoted to the culture of mathematics. I liked the periodical so much that I initiated a periodical for the culture of chemistry. It existed for 6 years. I did interviews with famous scientists for my publication, and not only with chemists. When the periodical folded due to acquisitions and mergers, I kept on doing the interviews and published my collection in six hefty volumes. There were at least a hundred Nobel laureates among our interviewees—in time my wife, Magdolna (in short, Magdi), and our son Balazs joined the project. A dozen of my interviewees received the Nobel Prize after, rather than before, the interview. For a while I was quite knowledgeable about who are doing the hottest science in chemistry, physics, biomedicine, and materials science. The development from the young student's love for mathematics to getting to know Nobel laureates appears to be a seamless story.

The other narrative demonstrates that seemingly blind chances at crucial stages of my life determined the outcome. I barely passed my first birthday when my lawyer father was killed as a slave laborer as a consequence of discriminative laws of anti-Semitic Hungary. I was not 3 years old yet when the train of cattle carriages started our journey to Auschwitz. By some fatal error, another train carrying a select group of people destined to survive had been sent to Auschwitz instead of Austria. Our train was to be a substitute; it moved some distance backward and went to Vienna. The *lager* (camp) was harsh and ruthless. My mother, my brother, and I survived. My grandmother did not. Upon our return to Budapest, we could not enter our house as it was now a Soviet command post. We established our new life deep in the country, my mother's birthplace. My new social status, my family's involvement in trade, relegated me to the class of capitalists —"class alien" was the awful label for it. It was the pretext to deny me the possibility of attending, first, high school and, later, university. At one point I was undesirable even for a locksmith apprentice of which there was an acute shortage. Still I ended up studying in high school and then at university. My studies cost a great deal for my parents and after 2 years at the University in Budapest, I signed up for study in Moscow, for which I was receiving a stipend. There, I learned not only good science but also to appreciate international interactions. After graduation, I created my line of research in Budapest, built up my group and laboratory, and became one of those at the top in my field. The most difficult and most rewarding was creating my experiments that were not a copy of anybody else's. I found an enthusiastic mechanical tool-maker whose talents had not been utilized and he was ready to unleash them. Because of our innovations, I could connect with a Norwegian group and then with Americans. My recognition abroad contributed to my election to the Hungarian Academy of Sciences. I entered the university sphere only after the political changes and was appointed professor in 1990. My whole career was characterized by independence, which I have valued from the start—it is a rare commodity.

To me the salient aspect of what I consider success has been the human relations. A high school mathematics teacher, a professor in Moscow, a Norwegian academician, a mechanical tool-maker, an editor of a book series, a painting by an impressionist artist, or a publisher, they all impacted my fate. I have selected a few dozen to introduce the reader to them. These brief introductions convey a flavor if not all the factual details. They make a mosaic from which the totality of my life does not come together, yet it suffices to create a sketch of my own portrait. Its message will depend on the interest, background, and motivation of each of my readers.

My mother gave me my compass. My wife has been my partner, friend, and muse, for more than 50 years. Our children, by now, grown-ups, have caused us limitless joy. They all hardly figure in the rest of this book, yet not a page, not a sentence, would have made it onto paper without them.

Acknowledgments

I prepared the Hungarian original of this book at the initiative of my friend László Tóth. Géza Komoróczy, Attila Paládi-Kovács, András Perczel, András Simonovits, Annamária Szőke, László Tóth, and Pál Venetianer helped it with comments and suggestions. The English version benefited from the careful critical reading of the manuscript by Annamária Szőke, Bob Weintraub, and Irwin Weintraub. My thanks to all. I am grateful for the encouragement received from Charlotte Hollingworth, Senior Editor, and Adelheid Duhm, Project Coordinator, Springer.

Budapest   István Hargittai
Spring 2019

Also by the Author

I. Hargittai, M. Hargittai, *Moscow Scientific: Memorials of a Research Empire* (World Scientific, 2019)

B. Hargittai, Ed., *Culture and Art of Scientific Discoveries: A selection of István Hargittai's writings* (Springer, 2019)

I. Hargittai, M. Hargittai, *New York Scientific: A Culture of Inquiry, Knowledge, and Learning* (Oxford University Press, 2017)

B. Hargittai, I. Hargittai, *Wisdom of the Martians of Science: In Their Own Words with Commentaries* (World Scientific, 2016)

I. Hargittai, M. Hargittai, *Budapest Scientific: A Guidebook* (Oxford University Press, 2015)

B. Hargittai, M. Hargittai, I. Hargittai, *Great Minds: Reflections of 111 Top Scientists* (Oxford University Press, 2014)

I. Hargittai, *Buried Glory: Portraits of Soviet Scientists* (Oxford University Press, 2013)

I. Hargittai, *Drive and Curiosity: What Fuels the Passion for Science* (Prometheus, 2011)

I. Hargittai, *Judging Edward Teller: A Closer Look at One of the Most Influential Scientists of the Twentieth Century* (Prometheus, 2010)

M. Hargittai, I. Hargittai, *Visual Symmetry* (World Scientific, 2009)

I. Hargittai, *The DNA Doctor: Candid Conversations with James D. Watson* (World Scientific, 2007)

I. Hargittai, *The Martians of Science: Five Physicists Who Changed the Twentieth Century* (Oxford University Press, 2006, 2008)

I. Hargittai, *Our Lives: Encounters of a Scientist* (Akadémiai Kiadó, 2004)

I. Hargittai, *The Road to Stockholm: Nobel Prizes, Science, and Scientists* (Oxford University Press, 2002, 2003)

B. Hargittai, I. Hargittai, M. Hargittai, *Candid Science I–VI: Conversations with Famous Scientists* (Imperial College Press, 2000–2006)

I. Hargittai, M. Hargittai, *In Our Own Image: Personal Symmetry in Discovery* (Kluwer/Plenum, 2000; Springer, 2012)

I. Hargittai, M. Hargittai, *Symmetry: A Unifying Concept* (Shelter, 1994; Random House, 1996)

R.J. Gillespie, I. Hargittai, *The VSEPR Model of Molecular Geometry* (Dover, 2012)

I. Hargittai, M. Hargittai, *Symmetry through the Eyes of a Chemist* (3rd Edition, Springer 2009, 2010)

Contents

1	Otto Bastiansen	1
2	László Bitó	5
3	Francis Crick	9
4	Mihály Csonkás	13
5	György Dalos	15
6	Edgar Degas	19
7	Lars Ernster	23
8	László Fejes Tóth	25
9	Árpád Furka	27
10	Pál Gadó	29
11	Richard L. Garwin	33
12	Ronald J. Gillespie	37
13	André Goodfriend	41
14	Árpád Göncz	45
15	József Hernádi	51
16	Avram Hershko	53
17	Lloyd Kahn	57
18	Gyorgy Kepes	61
19	Károly Kerti	65
20	George Klein	69
21	Arthur Koestler	73
22	Ferenc Lantos	77
23	Torvard C. Laurent	83
24	Paul Lauterbur and Peter Mansfield	89
25	Peter D. Lax	93
26	Sándor Lengyel	97
27	Alan L. Mackay	99
28	George Marx	103

29 Barbara Mez-Starck 107

30 Kurt Mislow 109

31 Charles T. Munger 113

32 Yuval Ne'eman 117

33 Paul Nurse 119

34 George A. Olah 121

35 István Orosz 125

36 Guy Ourisson 129

37 Michael Polanyi 133

38 Gabriela Radulescu 137

39 Andrei D. Sakharov 139

40 Géza Simonffy and Nikolai N. Semenov 143

41 Albert Szent-Györgyi 149

42 Edward Teller 153

43 Lev V. Vilkov 159

44 James D. Watson 163

45 Richard Wiegandt 167

46 Eugene P. Wigner 171

Epilogue 175

Index 179

About the Author

István Hargittai in June 2017 at a book launching in the Library of the Hungarian Academy of Sciences (photograph by and courtesy of Klára Láng)

István Hargittai is a physical chemist and Professor Emeritus (active) at the Budapest University of Technology and Economics. He is a member of the Hungarian Academy of Sciences and the Academia Europaea (London) and a foreign member of the Norwegian Academy of Science and Letters (Oslo). He is a PhD and DSc and has honorary doctorates from Lomonosov Moscow State University, the University of North Carolina, and the Russian Academy of Sciences. He is the Editor-in-Chief of the international periodical *Structural Chemistry*. He has authored and edited numerous books about structural chemistry, history of science, the nature of scientific discovery, memorials of scientists, conversations with famous scientists, and others. His books have appeared in English, Hungarian, Russian, German, Swedish, Italian, Japanese, Chinese, and the Farsi language. His wife, Magdolna Hargittai, is a physical chemist, honorary professor, and a member of the Hungarian Academy of Sciences and the Academia Europaea (London). She is a PhD and DSc and has an honorary doctorate from the University of North Carolina. Son Balazs is a professor of chemistry in the USA and daughter Eszter is a professor of communication studies in Switzerland.

Otto Bastiansen

I Learned from Him

Otto Bastiansen with the author in 1969 in Austin, Texas, and in 1982 during an outing in Norway (by unknown photographers).

It was a gray November day in 1967 in Budapest. I was staying at home with a broken leg in our sublet on József Boulevard. Magdi went to the airport to meet our guest, Professor Otto Bastiansen (1918–1995) of Oslo University. He was not our personal guest; rather, he was coming at the official invitation by the Hungarian Academy of Sciences. Magdi was a fourth-year student, majoring in chemistry at Eötvös University and I was a junior research associate in a research laboratory of the Academy. We had been married for 3 months and neither of us spoke good English. We had never spoken with Bastiansen, but I saw him at a meeting 1 year before. I could describe him to Magdi as a peripatetic man with an engaging smile.

I graduated from the Lomonosov Moscow State University (in short, Lomonosov University) in June 1965; and since August 1965, I worked at the Research Laboratory of Structural Chemistry of the Academy of Sciences. The Laboratory was located on the science campus of Eötvös University. In Moscow, I learned a technique of molecular structure determination that had not been practiced in Hungary. It was the gas-phase electron diffraction method. When I started my work in Budapest, I had to create all conditions for my research from scratch, both the experiment and the software for analyzing the measurements. I read everything in the literature, but the details of experiments were missing in the relevant papers. I turned to four scientists for further information; having selected those that I thought might be the most outstanding four in my field. Three of the four responded. The American professor kindly explained it to me why he thought I had embarked on an impossible task to create this area of research in Budapest, and he was *almost* proved right. The Japanese professor sent me four detailed research reports in Japanese. We had them translated and they proved to be useful. The Norwegian professor, it was Bastiansen, sent me some documentation, but even more essential, his letter was full of encouragement.

Bastiansen's visit was a great success. A crowd attended his lecture at the Academy of Sciences and his enthusiasm for science was contagious. He examined our experiments and spent time with us talking about them. He praised us, but it was also obvious that I needed additional training and he invited me for 3 months to his laboratory at Oslo University.

I arrived in Oslo on April 2, 1968. It was my first trip to the West and Norway was a fortunate place for such a first visit. Everybody and everything was friendly, including the English spoken by my Norwegian colleagues. I rarely saw Bastiansen at the University, because he was busy being its

© Springer Nature Switzerland AG 2020
I. Hargittai, *Mosaic of a Scientific Life*, https://doi.org/10.1007/978-3-030-34766-6_1

Rector (President) and the President of the Norwegian Research Council. But he gave me his free time generously. He lent me a bicycle and arranged for me a permission to use the university computer center day and night.

Bastiansen was already preparing for his 1-year visiting professorship at the Physics Department of the University of Texas at Austin, from September 1, 1968. He was a physicist, rather than a chemist by training, though this did not matter much in our field. Our kind of research was done mostly in chemistry departments in Europe and often in physics departments in the USA. He had an attractive offer from Texas. In addition to a high stipend, he could bring two of his associates with him to Austin. He had chosen two of his young Norwegian co-workers, and when one of them withdrew, he offered me the visiting position.

Bastiansen's American host, Harold P. Hanson (1921–2016), visited Oslo for a few weeks when I was there, and we spent some time together. He was the chairman of the Physics Department, which was huge, more like a big institute than a university department in Budapest or Oslo. He was ambitious and the University of Texas was on the rise. He convinced big names to join his department. He invited others for short visits. This is how in 1969 I could meet with the great Hungarian expatriate scientists Michael Polanyi and Eugene P. Wigner in Austin.

Hanson was considering bringing over a well-known physicist from Belgium, and he went to Brussels to discuss the matter. He finally refrained from inviting the Belgian scientist in spite of his interest to move to Texas. Hanson told me that the physicist was full of complaints about his current place and when someone complained so strongly about one place, he would complain about the next as well. This is a lesson to learn when one goes for an interview seeking a change in employment.

In Austin, I continued enjoying Hanson's company. Although he was busy, he made it a point to meet often for lunch with his immediate colleagues. Then he left Texas and moved to Gainesville, Florida, to become Dean of the Graduate School at the University of Florida, and then to Boston, to become Provost at Boston University. Finally, he served as the Executive Director, *Committee* on *Science, Space, and Technology, U. S. House of Representatives*. He retired in 1990. I visited him in every one of these new positions, and he visited us in Budapest. His hobby was translating Norwegian poetry into English—he was the son of immigrants from Norway.

Bastiansen started his visiting tenure in Austin on September 1, 1968, and so should have I. Alas, the process of receiving permission to go dragged on for months in Budapest. Finally, I could leave in January, 1969. Bastiansen and I overlapped for half a year in Texas. We shared an office, also with Jon Brunvoll, Bastiansen's other associate. I enjoyed the constant stream of Bastiansen's visitors. He was full of ideas and stories. He had interesting chats with our African-American janitor, a Mr. Pemberton, who had a degree in English literature. Bastiansen introduced me to Ilya Prigogine, the Russian-Belgian scientist, who in a few years' time would receive a chemistry Nobel Prize. Neither Jon, nor I could drive, and Bastiansen taught us with great patience and a great deal of scientific explanations. I still remember his instructions when I am driving onto an expressway or negotiating a 270-degree turn.

Our interactions lasted to the end of his days. When in 1978 my brother stayed abroad "illegally," and I was expecting bullying by the Hungarian authorities, he gave me advice. He told me to never volunteer any information unless asked, not even a remark about the nice weather. Even the most innocent comment may lead to conversations that could hurt me, he said. He was experienced in being interrogated at the time of the German occupation of Norway. I followed his advice. I was quarantined for a few years and the official letter declining my application for a passport did not hide the fact that I was penalized because of my brother's action. When there was a scientific meeting in celebration of Bastiansen's 60th birthday in Oslo, I was an invited presenter, but was not allowed to go. My lecture that was not delivered was published in a Norwegian journal and my absence received greater publicity than my presence would have.

My travel ban was still in effect in 1981 when Oslo University selected me for the annual Hassel Lectureship. The lecture had to be postponed repeatedly because of my difficulties in obtaining permission to attend the event. I was not denied, but I was not permitted either. Finally, I had my passport. Odd Hassel (1897–1981) was the great old man of Norwegian science. He received the Nobel Prize in Chemistry in 1969. When I was in Oslo in 1968, he was kind to me and gave me two valuable books when I was leaving. He did his most important research during the German occupation, but refused to publish his results in German. He was not allowed to publish in English, so he published in Norwegian. After the war, he republished his essential achievements in English. His Nobel Prize in 1969 was awarded for discoveries he had communicated in Norwegian during the war. These were papers that many people cited, but hardly anybody read.

My Hassel Lecture was announced for May 11, 1981. Hassel was expected to attend, but died at dawn on May 11. The leaders of the University faced the dilemma of what to do with the Hassel Lecture. It was at that moment that Bastiansen remembered the day when the Gestapo arrested Hassel. There was this dilemma whether the University should declare strike in protest, but Hassel sent word from the prison, "The lectures must go on." This is what was then decided for my Hassel Lecture. Due to the sad circumstances, my Hassel presentation received much press coverage. It was appreciated that before I embarked on the technical details of our work, I included a summary of Hassel's achievements. My Hassel Lecture was perceived as a proper memorial to the great scientist. Bastiansen gave me a Victor Vasarely (see in the Orosz chapter) silk print he had received from the artist years before. The Norwegian Academy of Science and Letters elected me foreign member in 1988 and the ceremonial inauguration was attended by King Olaf V, the honorary President of the Academy. Olaf V received me for an audience following the inauguration and Bastiansen, the President of the Academy, was also present; he and the King were old friends.

I learned a great deal from Bastiansen and intense cooperation with his associates followed for many years. Jon Brunvoll with whom I overlapped in Austin spent twice his 1-year sabbaticals in our laboratory in Budapest. Others came for shorter and longer stays. Over thirty joint Hungarian–Norwegian publications came about from our work. I saw Bastiansen for the last time in an assisted living facility where he had moved after retirement and he did not want to burden his three daughters. I went to see him with one of his former associates. He was lonely but came to life during our visit. We covered a wide range of topics in our conversation and when we wanted to leave sensing him getting tired, he did not let us. At the end, as we were leaving, he hugged me—something he had never done before.

László Bitó

Charitable Xalatan

László Bitó autographing his books in 2014 (photograph by the author).

"Charitable Xalatan" refers to the benefits of this eye drop to millions of people suffering from glaucoma and to the charities László Bitó (1934–) has performed from his income from Xalatan. He had an extraordinary career. During the dark Soviet-style Rákosi dictatorship in the 1950s in Hungary, he was excluded from studying and was forced to do slave labor in a mine. His dream to become a writer was shattered. He took some active role in the 1956 Revolution and when it was brutally suppressed, he became one of the over two hundred thousand refugees fleeing the country. He landed in the USA where he could study. Becoming a writer was not in the cards there either as he could hardly speak English. His love of reading and his interest in vision moved him toward ophthalmology—this may have been

his real motivation, or this notion may have also developed later on.

Bitó received a scholarship from Bard College in Annandale-on-Hudson, New York. He majored in chemistry with a minor in biology, and graduated in 1960. He did graduate work at Columbia University and became a PhD in biophysics in 1963. Following postdoctoral studies at the University of Louisville in Kentucky and University College London, he returned to Columbia University as a research associate of the Department of Ophthalmology. He retired in 1997 as Professor of Ocular Physiology and in 1998 he became Professor Emeritus of Columbia University. Upon his retirement, he moved to Budapest and during the past 20 years, he finally fulfilled his childhood dream of becoming a writer. He is an active participant in societal life, worries about the destruction of democracy and supports pro-democracy movements. Most people know him for his charitable activities and are not aware of his shining career in science.

My wife and I met László Bitó in 2007 in the home of Endre Balazs[1] in Fort Lee, New Jersey, just a bridge (the George Washington Bridge) away from Manhattan. There will be more about Balazs, another Professor Emeritus of Columbia University, in the second half of this chapter. When I met Bitó for the first time, I was not familiar with his literary activities, but knew a little about his science. This is what I am going to describe here in a nontechnical way.

Bitó focused his research from the start on glaucoma. The most characteristic syndrome of this condition is the elevated internal eye pressure, which is a consequence of the unbalanced production and elimination of a fluid, the vitreous humor, in the eye. Bitó wanted to try the application of the prostaglandin hormones to stimulate the elimination of the fluid. Other scientists had already tested prostaglandins and concluded that they are not suitable for such treatment: They tried large amounts of prostaglandins. This is a standard approach; if large amounts cause no harm, it is supposed that smaller amounts would have no negative consequences either. The application of large amounts in the eyes of the test animals, however, showed severe toxicity. This stopped further experimentation, but Bitó was an exception. He reasoned that even if large amounts showed toxicity, small amounts may not. This sounds logical, but at the time the idea appeared out of the ordinary and invited sharp criticism.

Fortunately, Bitó persevered. The prostaglandins are hormones, and hormones perform their vital functions in very small amounts. Thus, the experience based on the application of large amounts should not be decisive. His other consideration was the diversity of the eyes in different animals. He suggested that the final conclusions should not be drawn from experience conducted on one type of animal eyes only. Previous experiments used rabbits that usually fall prey to other animals and whose eyes protrude to enable them to see everything around them. In contrast, the eyes of predatory animals, such as cats and monkeys, developed sharp focus; they are deep seated and well protected. Bitó

found that the eyes of cats, starting with his own cat, and his own eyes, well tolerated the administration of small amounts of prostaglandins and he observed the decrease of internal eye pressure. He worked out techniques of measuring the pressure in the eye in less invasive ways than others used to do. Years of painstaking research bore fruit and Bitó came up with an eye drop, which he called Xalatan that helps alleviate the suffering of glaucoma patients.

Bitó did not want to deal with the commercialization of his discovery and let the appropriate office of Columbia University handle it. He filed for a patent though, on May 3, 1982, which was accepted on July 8, 1986. The gap in the dates indicates that it was not a trivial process. When it was first suggested to Bitó to file for patent, he was surprised. He was familiar with Edison's activities, but did not know much more about patenting. He did not think discoveries in biology could be patented as the Creator had already held patents on everything. The Columbia office was not successful in finding an American company that would have undertaken the utilization of Bitó's discovery. When 10 months had passed already, Bitó turned to Endre Balazs for help. At the time, from 1975, Balazs was the Malcolm P. Aldrich Research Professor and Director of Research at the Department of Ophthalmology.

It was an unusual coincidence that there were two Hungarian expatriates working as research professors at the Department of Ophthalmology of Columbia University. Both brought world fame to their host institution. Balazs had built up fruitful interactions with a Swedish pharmaceutical company, Pharmacia, and now he connected Bitó with this company. There were plenty of difficulties. For example, in one trial, there occurred skin coloration and they suspected melanoma. Only one such case would have destroyed any further hope for the application of the eye drop. Fortunately, something else caused the color change. In 1996, the Food and Drug Administration (FDA) issued permission for marketing Xalatan, and it has become a big success. By 2003, the earnings have reached one billion US Dollars. Eventually, several companies joined the market of the eye drop upon the expiration of the original patents and it is no longer possible to follow the earnings reliably. Bitó received a small fraction of the earnings, which meant a relatively large income for him.

[1] Balazs was his last name; more often, Balazs is a first name in Hungary.

László Bitó, Endre Balazs, and the author in 2007 in the Balazs home in Fort Lee, New Jersey (photograph by Magdolna Hargittai).

The interaction between Bitó and Balazs dated back years and Bitó's association with Columbia University had started long before Balazs had joined it. When the position of research director at the Department of Ophthalmology became vacant, Bitó suggested Balazs for the job. At that time, Balazs worked in Boston, but found moving to New York attractive. His interest and qualifications eminently fitted the position of the research director in an ophthalmological institution.

We had known Balazs since 1999 and met him through Torvard Laurent (Chap. 23). We had several brief meetings with Balazs between 1999 and 2007. Then, from 2007, for seven years we spent a few months annually in his Matrix Biology Institute in Edgewater, New Jersey. We were looking into the molecular structure of hyaluronic acid and dealt with various aspects of the history of research on this substance. Some of the work was done in the popular resort, Saint-Tropez, France, the summer residence of Balazs and his wife.

Endre A. Balazs (1920–2015) was born in Budapest and attended the Fáy András Gimnázium between 1929 and 1937. Throughout his high school years, he wrote essays on social science and natural science and was thinking about becoming a politician. Nonetheless, he enrolled in medical school in Budapest and became an MD in 1943. As a freshman, he joined Professor Tivadar Huzella for extracurricular activities doing research on intercellular substances.

Eventually, his interest narrowed to hyaluronic acid, which became his lifelong project. The aqueous solution of hyaluronic acid is a dense, viscous liquid, very similar to egg white. The liquids in the joints, such as the knee, contain large amounts of hyaluronic acid ensuring flexibility. In his American career, Balazs became the owner of exceptionally successful patents for the application of hyaluronic acid in medicine, including the treatment of joints. However, he filed for his first patent back in Budapest in 1943. He collected joint liquid at slaughterhouses, added sugar to the liquid and whipped it, and thus produced meringue without an egg. It could have become popular in wartime in view of the severe egg shortage, but nobody ever utilized it.

From the early 1940s, Balazs started thinking about emigration. The signs of the emerging communist dictatorship in the post-war years strengthened his intention. He had a good position at the medical school for doing research, and this made it possible for him to join a delegation to attend the International Congress of Cell Research convened in 1947 in Stockholm. He held a well-received talk at the meeting, followed by job offers. He chose one at the Karolinska Institute. There he conducted research and when a young student joined him, Torvard Laurent, it was the beginning of a lifelong association.

In 1950, in part to advance his research career and in part of being afraid of the closeness of the Soviet Union, Balazs and his young family immigrated to the USA. From 1951, he was busy in organizing the Retina Foundation in Boston for conducting ophthalmological research, and he continued his work on joints as well. He noticed that the hyaluronic acid molecules of large molecular weight may break into molecules of smaller molecular weight, especially with aging. This reduction of molecular weight is accompanied by pathological alterations in the joints. Balazs founded his first company in 1968, a miniscule one, with the purpose of producing pure hyaluronic acid that could then be used for medical purposes. Hyaluronic acid had been used in ophthalmic surgery, but widespread application became possible only when Balazs had worked out its purification.

When Balazs retired from Columbia University in 1985, he founded the company Biomatrix for producing and marketing hyaluronic acid. Thus, he created a full chain of innovation from conducting basic research to delivering the product ready for the shelves of stores or for its use in surgery. He continued his activities after his second retirement in 2000: he founded and directed the Matrix Biology Institute until the end of his life.

One day, there may be a film about his life and work. Just imagine the scene when a racehorse was injured and could not stand on its legs. A single injection of hyaluronic acid in its joints let it stand up and the horse soon started to race again. Or the man who could not see since his childhood and nobody would operate on him because the retina in his eye could not be protected during the surgery. When hyaluronic acid had become available, its introduction in the eye protected the retina, the operation was successful, and the man could see again!

The principal source of hyaluronic acid used to be the cockscomb and it was crucial to apply the purification technology Balazs had worked out. The purity had to be monitored and it was done by administering drops to test animals. In order to find the appropriate monkey for the purity check, Balazs examined numerous monkeys around the world and his research enriched the scientific literature. Today, it is no longer necessary to use test animals, because pure hyaluronic acid is produced by bacteria in large quantities. At an early stage of Balazs's career, he was offered a large sum for the rights of his technology. The offer would have sufficed freeing him from financial worries for the rest of his life. He declined and continued his activities that involved fresh starts, risks, hard work, creating new laboratories and new research groups, and financial sacrifices, let alone the legal battles when competitors tried to question his priorities. Balazs prevailed; he never gave up.

Scientists' Scientist

Francis Crick in 2004 in the Cricks' home in La Jolla, California (photograph by the author).

The names of James D. Watson (1928–) and Francis Crick (1916–2004) are linked forever in the annals of science. They were the co-discoverers of the double-helical structure of deoxyribonucleic acid, DNA, the molecule of heredity. Both then stayed at the top of science for decades even if for different reasons: Watson (there is a separate chapter about him) as a science administrator and as an author; Crick, as a thinker and inspiration for new vistas in research. Their personalities were vastly different. In our long-term project of interviewing famous scientists, I often asked my interviewees about their role models. Many named Crick; none named Watson.

I started corresponding with Crick in 2000 and our exchanges lasted to the end of his life. We met only once when on February 8, 2004, Magdi and I visited the Cricks in La Jolla. Crick died a few months later, on July 28. There was no chitchat in our correspondence. Usually I posed questions and Crick responded. He took all my questions seriously although he was famous for brushing off people who tried to correspond with him. In our first exchange, I raised questions related to crystallography and about his joint work with Sydney Brenner (1927–2019), the South African-born British biologist. We corresponded about scientists who might be Crick's disciples and about his excellence in lecturing. I asked him about what he considered to be success in science; whether Rosalind Franklin (1920–1958) might have known that Watson and Crick had had access to her results in the X-ray crystallography of DNA; and about his evaluation of George Gamow's (1904–1968) contribution to science. The Russian-born American physicist Gamow proposed the Big Bang model for the origin of the Universe. He was also the one who as soon as Watson and Crick published their suggestion of the double-helical structure, raised the question of the genetic code (discussed later in this chapter). I was also curious of Crick's views on religion.

There is an annual Rickman Godlee Lecture of great prestige at University College London. Crick gave it in 1968, entitled "The Social Impact of Biology." In his presentation, he focused on the emergence of modern biology, especially, genetics, and of biotechnology, and on how they changed our perception about our place in the Universe. He elaborated on their effects on a variety of fundamental issues, such as demography, medicine, economics, and societal problems, and called for constructing a new ethical framework based on science rather than religion. In my extensive conversations with Watson in 2000, he mentioned two issues from Crick's Godlee Lecture, and I raised them in my correspondence with Crick. In 1968, Crick said that a newborn baby should only be declared alive, 2 days after birth. He also said that the state should not spend money on medical care of people above 80.

Crick responded that he did indeed give a provocative lecture in 1968 at University College London. Today (this was in 2001), he would modify his suggestions. In the old days, doctors quickly let a very deformed or handicapped baby die, rather than make exceptional efforts, as they often do now, to keep the baby alive. Thirty years later, Crick no longer thought his suggestion realistic, especially not in the USA, to count life as starting after the first 2 days of a baby's life. Many religious people believe that life starts much earlier, even at conception. In his modified view, Crick thought that one has to consider not just the feelings of the baby but also the feelings of the parents and other members of society, however silly one may think them to be—he added. Still, Crick believed that doctors should not make exceptional efforts to keep a very handicapped baby alive.

As to the age limit, Crick realized that people now live longer than they did in the 1960s, so he would push such an age a little higher, and he doubted if a rigid rule would be acceptable. He thought that very expensive treatments, or ones that have only limited availability, should be allocated in some sensible way. He mentioned the State of Oregon where such a scheme was under consideration. Crick was 52 at the time of his Godlee Lecture and he was 84 at the time of this exchange and just after a heavy surgery. He also stressed the right of a person who is incurably ill to terminate his own life.

In 2001, I was completing my book about the Nobel Prize, *The Road to Stockholm*, and devoted a chapter to missing Nobel Prizes. I wondered about Sydney Brenner's absence in the roster of Nobel laureates. I knew that Brenner and Crick used to work together in Cambridge and that they had a sizzling intellectual interaction for years. It seemed to me that from the point of view of the Nobel Prize, theirs was an asymmetric relationship. Whereas Crick had already had his award, Brenner's close work relationship with Crick might have hindered the assignment of any major research achievement to Brenner. Crick responded that Brenner and he shared an office for 20 years, but for most of the time Brenner was doing experiments in the laboratory. They did talk together for an hour or more on most days. He opined that Brenner ought to have the Nobel Prize. Awarding him the Prize may be complicated by Brenner's vast output of significant work that made it difficult to select just one particular discovery for the Nobel distinction. Crick added that Brenner's work was widely recognized and that Brenner

has received every other award other than the Nobel—many more than Crick has! Our exchange happened 18 months before Brenner's long awaited Nobel Prize was announced in October 2002.

Crick had partnerships with other scientists, but always with one at a time. I asked him whether there were any scientists who could be considered directly his pupils. I found this an intriguing question in light of so many scientists naming him their role model. However, Crick could not name anyone as his pupil; he supervised a graduate student for 1 year, but after that someone else took the student over. He may have deliberately avoided such tasks. For major collaborators, he named Watson, Brenner and (more recently) Christof Koch (1956–). For transient collaborators, he named Aaron Klug, Beatrice Magdoff, Leslie Orgel, and Graeme Mitchison. His collaborators then had numerous pupils.

As for religion, Crick considered himself agnostic leaning toward atheism. He referred to Pope John Paul II who considered evolution as a fact, but condemned Crick—this is how Crick saw it—for his activities. Crick was saddened seeing the many millions of people in the USA who still discard the theory of evolution and think that the Earth is not yet ten thousand years old.

When we were planning our trip to California for January–February 2004, I wrote to Crick that we would be on a brief visit at the California Institute of Technology in Pasadena. I was not suggesting anything because I knew about his illness. He wrote back that being so close it would be a shame not to come and see them in La Jolla. Still he asked to check with his assistant on the eve of our visit, because his condition may turn critical at any moment. Nothing happened and at the appointed time we rang the bell at the Cricks' residence. The door opened, and there he was, "I am Francis Crick," he said.

It was a great visit. Odile Crick gave us salmon for lunch, my favorite. A great variety of topics came up in the conversation. When we left, Magdi and I each recorded everything what we remembered from the visit, right away.

Of the scientific topics, I single out one. It is not unambiguous from the literature who was the first to raise the issue of the genetic code. Let us start with defining what the genetic code is so that we all agree what is what we are talking about. The genetic code is the means by which the information is transmitted from the bases of the nucleic acids to the amino acids of the proteins. Crick agreed with us that the story needed clarification for he had discussed it in presentations only without ever having recorded it in any publication. Even his lecture notes have been lost. Hence, I feel it necessary to convey what he had to say in this connection.

The transformation of information in question is from the one-dimensional sequence of the nucleic acid to the three-dimensional structure of the protein. This means that the one-dimensional sequence of the nucleic acid determines the folding of the protein molecule. The essence of Crick's reasoning was that the replication of the sequence of the nucleic acid suffices for the complete replication. So who raised the question about the way this is accomplished, i.e., who first raised the question of the genetic code?

According to Crick, he and Watson as early as spring 1953 considered this question. They could announce that they had solved the secret of life on February 28, 1953, at the Eagle Pub, in Cambridge, UK, because they had recognized this relationship. Merely the double-helical structure of DNA would have not sufficed for making such an announcement. They had been considering the question of information transfer for some time and this enabled them to see the weight of their discovery for genetics. The recognition of the significance of the sequence preceded the discovery of the structure, and this was due to Crick. The structure of DNA and everything they produced in relationship to the genetic code was their joint result. In this, their achievements could not be distinguished.

The origin of the list of the 20 naturally occurring amino acids also came up in the conversation. It was Crick and Watson who compiled this list for the first time. This is seldom mentioned though it is a result not to be ignored, Crick added. There are scientific findings that do not appear as important at the moment of the discovery as they become in time. The compilation of the list of the 20 amino acids could be associated more closely with Crick and Watson than it is. However, the discovery of the double-helical structure of DNA so soon relegated the compilation of the list to the background of events.

All four of us participated in the conversation with Crick leading it. At one point, Odile's contribution was rather moving. The discussion was about the attitude of the scientist who is interested only in producing new knowledge, while remaining oblivious of priority considerations. Leo Szilard, George Gamow, and J. Desmond Bernal were our examples. At this point Odile interjected that Francis was also such a scientist.

We asked Crick about Szilard something we had already asked of Watson and of the French Nobel laureate François Jacob. It was always difficult for Szilard to find a job suitable for his personality. The question was whether Crick would employ Szilard (or somebody like Szilard) if he would suddenly appear at his doorstep? Watson thought it was improbable that he could offer Szilard employment because Szilard was always two steps ahead of everybody else. In contrast, Jacob would have gladly employed Szilard exactly because Szilard was always two steps ahead. His job could be a special one of a bumblebee, a one-person communication system, talking with people and disseminating news. Crick agreed with Jacob, but he understood that it is not easy to arrange support for such special scientists as Szilard. On the other hand, such support would be negligible as compared with the enormous expenses of contemporary research projects.

Odile and Francis Crick (photograph by the author).

There was no end to the exciting topics in our conversation, but at some point we realized that we should take our leave. We were impressed by the adventure of this meeting and grateful to the Cricks for letting it happen. It was a comforting feeling that they seemed to enjoy the day we spent together; it may have meant a relaxation from the routine of the serious illness. The weight of Crick's condition was not perceived during our get together, but he succumbed to the illness in less than 6 months. In our memory, he stayed as someone with a keen interest and one who appreciates humor—I am still hearing his legendary uproarious laughter.

Mihály Csonkás

From the Ancient to the Present

Szent László Gimnázium, 28–32 Kőrösi Csoma Sándor Avenue, Budapest (photograph by the author)

Just a few days before the start of the academic year 1955/56, we received a note from Táncsics Gimnázium—the local high school—in Orosháza, where we lived, a town far away from Budapest, in Southeastern Hungary, a rather impoverished region. Before World War II, it was "the largest Hungarian village. Its size sufficed for being a town, but the local leadership prevented the upgrade to avoid higher taxes. It became a town in 1949 without any change in its size. The note from the high school was about the annulment of my enrollment in the school. The justification from the point of view of the communist regime was that my maternal grandfather (who died in 1942) used to be a shop owner. It took a 6-week struggle for Mother before the Ministry of Education determined that injustice happened. The Ministry decided that I could become a student in any high school in the country, except where we lived. They did not want to diminish the authority of the local organs of the communist party and government. I chose a school in an outer district of Budapest because my brother was hoping to receive an apartment there (he did not). I found temporary accommodation with a cousin and her family in their small home. I had to travel every morning and afternoon three times changing trams. Often, I was hanging on the stairs of the crowded tram—this was long before automated closing doors. The school was the László I Gimnázium; before and now again Szt. (Saint) László. The Principal knew about the reason why I started 6 weeks late, and he was kind to me. My homeroom teacher, Dr. Mihály Csonkás, did not know. When I began with failing grades, he suspected that I gained acceptance in the school through connections and was not fit to study. He told the class about his suspicion.

My start was truly difficult. I had to organize my life in Budapest and could not study during the first days. My situation was made worse by my lacking grades. I was asked for recitation more than the other students, so the failing grades accumulated rapidly. I was devastated. Here I finally had the opportunity to study and was failing. My classmates saved me, especially the best student. His name

© Springer Nature Switzerland AG 2020
I. Hargittai, *Mosaic of a Scientific Life*, https://doi.org/10.1007/978-3-030-34766-6_4

was Gyula Csocsán and our good relationship continued even when I could be counted as his rival. Years later, I tried to find him; but never succeeded. I still appreciate the magnanimity of this 14-year-old boy.

Eventually Dr. Csonkás changed his opinion about me and valued my contribution. It was not a strong class; many of the students commuted from nearby villages. They enrolled with shining recommendations that could not mask their lack of a good basic education, and Dr. Csonkás was very demanding. He taught history and Hungarian language and literature. There was homework every day. One day we had to write 30 lines about the fall foliage; the next day 20 lines, and the next day 10 lines. The fewer the lines the harder it was.

First grade history was the times of the ancient Greeks and Romans. Dr. Csonkás enjoyed lecturing about them. Those ancient stories came alive in his presentation and he related his stories to universal human values. He gave a memorable lecture about the Greek Thucydides, who lived more than 400 years before the Christian era. Thucydides became a history writer after the great Athenian statesman Pericles had defeated him in a power struggle. We owe it to Thucydides that we know so much about the golden age of Athens and the deeds of Pericles. Through Thucydides's eyes, Pericles is seen as a great statesman and military leader. He could have written disparagingly about Pericles, having suffered a defeat at his hands, but, Dr. Csonkás taught us, by belittling Pericles, Thucydides would have belittled himself. It is more honorable to have been defeated by a great person than by a nobody. It was a life lesson.

Close to the end of the academic year, Dr. Csonkás disappeared. He may have been transferred to another school or may have retired. That he was an extraordinary teacher I understood only when we met his replacement. He was correct and knowledgeable, but certainly, he was far from

extraordinary—he was no Dr. Csonkás. Upon the completion of my freshman year I also left this gimnázium.

I was already a university student when I visited Dr. Csonkás. It bothered me that we had no opportunity to bid goodbye to him. Also, I was increasingly appreciative of all he taught us. He lived in District VIII, one of the poorer inner city areas. He was visibly happy to see me and I was sure happy to see him.

For my second year, there was no reason for me to attend school in such a faraway district, so I moved to Kölcsey Gimnázium, which was close to the place where Mother had found a sublet room for me with breakfast. It used to be the no longer existing domestic help's room in the comfortable apartment of two spinsters. The rent was stiff and Mother had to send them a food package twice a month under the notion that food from the "country" was what they wanted to have instead of the stuff they could buy in the store. In reality, it was just an addition to the rent. Also, I had to have cocoa and cake every Sunday afternoon with them and their niece who came to visit. She was a nice girl who went to an all-girls school and her great-aunts wanted to help her get used to the company of boys.

This arrangement lasted from September 1 to October 23, the day of the 1956 anti-Soviet Revolution. Instructions stopped and every day during the next 10 days I was touring the city. On November 4, the Soviet tanks brutally ended everything. In the morning of November 11, I started my journey home. I had to hitchhike—there were no trains or buses—and I was lucky to arrive home in the evening of the same day. When, months later, I returned to my former landladies for my affairs, I found everything intact, except for two things. My collection of insects and my collection of revolutionary newspapers disappeared.

One Sentence Suffices

György Dalos in 1964 (photograph by the author).

György Dalos (1943–) and I had three overlapping years (1962–1965) as students at Lomonosov University. For that period, we were best friends. Everybody is familiar with this flagship institution of higher education in Moscow, and its characteristic tower. World War II hardly ended when the megalomaniac Soviet dictator, Joseph Stalin, decided to have seven skyscrapers built in Moscow. He thought that Moscow could not be a bona fide world capital without skyscrapers. Construction started before even each of the seven had been assigned their function. The Rector (President) of Lomonosov University, Aleksandr Nesmeyanov, petitioned the building designed for Lenin Hills (before and now again Sparrow Hills) for his School.[1] The University had long outgrown its old campus in downtown Moscow. His request was fulfilled.

Dalos studied history and I studied chemistry. There was a mixed crowd of Hungarian students at Lomonosov University. The father of one of the students was a cabinet member back home, and everybody knew about him. Nobody knew anything about Dalos's parents or mine. His father became ill in a slave labor camp and died soon after liberation in 1945. His mother was often ill and Dalos spent much of his time in an orphanage. He had a good sense of humor, a very tolerant nature, but stubborn as far as maintaining his humanism. He was a loyal friend and wrote poetry. We spent quite some time together but even more with our new Russian friends whereas for Hungarian students it was customary to seek out each other's company. There were about 50 Hungarian students at the Lomonosov at the time, both undergraduate and master's. There was then a separate group of what we would call today, doctoral students, and they were considerably older than today's doctoral students.

There were fierce debates within the Hungarian community, and Dalos and I were usually in the middle of the often-extreme positions. I graduated 2 years before him. There was a big discussion when I was getting close to graduation with my master's degree about my character reference to be sent back to the Ministry of Education. Some wanted to include a

---

[1] Although Lenin Hills has regained its original name, Sparrow Hills, the official address of Lomonosov University has stayed 1 Lenin Hills. In Russia, old names were returned, but where there was no original name, the names of Soviet time have stayed. Thus, for example, Leningrad was changed back to Saint Petersburg, but Saint Petersburg is still in the Leningrad Region.

© Springer Nature Switzerland AG 2020

I. Hargittai, *Mosaic of a Scientific Life*, https://doi.org/10.1007/978-3-030-34766-6_5

negative comment about my inclination for compromise in debates whereas others wanted to protect me from this damning "accusation." I did not really care although I understood that at the time accepting compromises was condemnable. I forget whether it finally figured in my character reference or not (it probably did).

Béla Uitz in 1969 in his studio/home in Moscow (photograph by the author).

Dalos did some cultural reporting for Budapest magazines on the side. I went with him when he visited Béla Uitz (1887–1972), a former avant-garde artist who escaped the wrath of the white terror of the early Horthy era and immigrated to Moscow. During his four decades in the Soviet Union, he became an artist of monumental frescos in the so-called socialist realist style. He told us about his mural at a permanent exhibition of economy, which he had to remake. The censors determined that the trucks and the people were moving away from, rather than toward, the future. I took many snapshots of Uitz and years later gave them to a museum in Pécs (Magdi's birthplace) that collected Uitz's art.

I visited another Hungarian expatriate in Moscow, Jenő Varga (1879–1964), a close relative of one of my fellow students in Budapest. Varga was a minister of finance of the short-lived Hungarian Soviet Republic in 1919. He was in great repute as an economist in the Soviet Union. Many of his comrades had fallen victim to Stalin's terror, but he survived. He lived in an upper-floor apartment on Leninsky Avenue, a choice location. He showed me the view from his living room and one could see far away although the view was dominated by rooftops. He told me that I would never see such a beautiful view in Budapest—he was biased to this extent. When I was in Moscow recently, I noticed a memorial plaque on the façade of his former home (11 Leninsky Avenue) recording that Evgeny Samuilovich Varga, noted fighter of the international communist movement, full member of the Science Academy, used to live and work here.

Dalos's first book was published in Budapest during our student years. It was a volume of poetry, titled "Birth of Our Words" (*Szavaink születése*). He later became a renowned writer in Berlin. Nowadays we seldom meet, but I can always turn to him when I need some piece of poetry in Hungarian translation.

Dalos reminds me of poetry. What I like best about poetry is that a poem can express even complex thoughts, ideas, feelings, situations, and phenomena in a most succinct way. As I started recording these lines in late fall 2017, there was yet another reason why I remembered poetry. I just read Timothy Snyder's new small book, *On Tyranny: Twenty Lessons from the Twentieth Century*. I had a feeling of déjà vu as I was progressing in the book. It is about the looming danger of totalitarianism as American society is navigating through Donald Trump's presidency. My life began under the Horthy regime with ever more horrifying anti-Jewish legislation culminating in our deportation. We survived the camp and, following a few years of fragile democracy, we lived for 40 years under Soviet-type socialist rule. In the 2010s, we are reliving a gradual destruction of democracy and the formation of an autocratic order. However, this was not the only reason for my déjà vu.

It was a feeling while reading Snyder's *Tyranny* that I have already read it, the same, yet not quite the same. Then it dawned on me that Snyder's *Tyranny* brought back the Hungarian poet Gyula Illyés's poem, "One sentence about tyranny." Snyder narrates about the potential omnipresence of tyranny and this is what makes the poem "One sentence" so frightening. Tyranny is not only what we readily associate with it, the violence and the ruthless suppression of the most diverse rights. Rather, it manifests itself even in the most subtle, the most intimate personal aspects of our daily lives. Even in our death, we still remain its captive. Illyés wrote his "One sentence" in 1951, in the darkest time of the communist dictatorship when he could not hope it getting ever published. It was first published in 1956 during the few days of the anti-Soviet revolution. When the revolution was crushed, the poem was forced back underground for another 30 years. The worst feature of tyranny was that at some point there was no longer any need for it to spell out all restrictions and barriers. The attitude of people embraced even the unspoken debilitating rules and they no longer needed to be told how to behave. Everybody and everything complied fulfilling what even was only supposed to be the expectations of tyranny. This frightening aspect of an autocratic regime in the making made Snyder compose his *Twenty Lessons*.

In 1978, my brother stayed out of Hungary "illegally." Soon I applied for my passport as I was invited to give a talk in Norway. My passport was denied, that is, I was not allowed to travel, and the reason was given, officially, that this was because my brother was abroad illegally. Punishing family members for the "crime" of another family member was a Nazi approach, but this did not bother the authorities. This was also what I expected. What I did not anticipate was the attitude of some of my friends. One of them called me and told me that my brother's act would not change anything in

our friendship, and this was the last I heard from him, ever. He was a fairly high-ranking official in foreign trade, so his fear was understandable. Another friend was more honest. He told me that for a while we should not meet, for obvious reasons, he added. He was an oil engineer in a rather low position; nonetheless, he was afraid. In my workplace, a research institute of the Science Academy, the director and his deputy told me that for a while I should exercise restraint. They did not spell out what they meant, but the atmosphere of the encounter was chilling. It made me feel as if I had become infested with a terrible disease, highly contagious. One of Illyés's stanzas expresses what I could only describe in many more words, that tyranny was there, everywhere[2]:

> in the street,
>   in the routinely repeated how-are you-s,
>   in the handshake
>   that suddenly loosens, . . .

It is unfortunate that none of the translations, and there are several, can express exactly the exasperation and desperation I deeply feel when reading this poem.

---

[2] Illyés Gyula (Gyula Illyés): *29 vers. 29 Poems* (In Hungarian and in English, Budapest: Maecenas, 1996), pp. 34–49.

Learning from the Arts

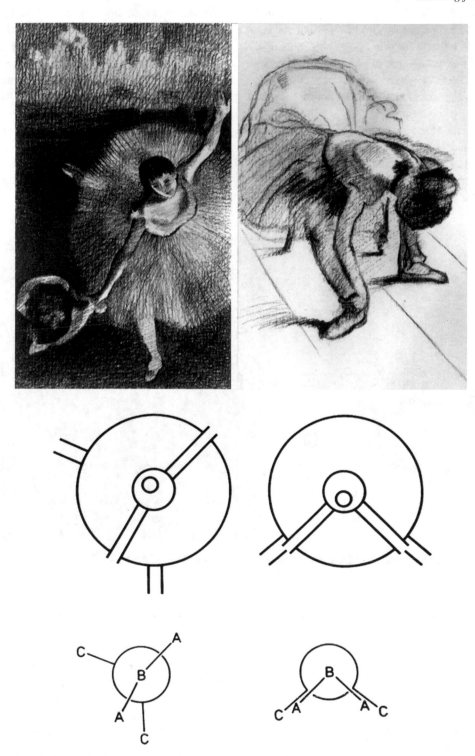

Degas' dancers in the sketches by the late Ferenc Lantos illustrating rotational isomerism (called "Newman projections" in the scientific literature).

© Springer Nature Switzerland AG 2020

19

I. Hargittai, *Mosaic of a Scientific Life*, https://doi.org/10.1007/978-3-030-34766-6_6

Impressionism was my first favorite art period to which I have added more modern directions. The impressionists enthrall me to this day. Edgar Degas (1834–1917) provided me with an experience for the first time in recognizing a helpful analogy between a piece of art and a scientific concept. We were at the Musée d'Orsay in Paris when I had an overwhelming déjà vu in front of Degas' "End of the Arabesque." I felt as if I had seen this image before though I knew I did not, but it reminded me of something not obvious immediately. Then, suddenly, it came to me. The image resembled one of the rotational isomers of a chemical structure.

Imagine a molecule of the formula $A_2B–BC_2$. There is a B–B bond in the center, and there are two other atoms linked to each of the two B atoms. Imagine now the B–B bond to be the dancer's body and her skirt representing a plane bisecting the B–B bond. Let the B–A bonds represent the arms and the

isomer, I soon found another Degas for the eclipsed isomer in the Hermitage in Leningrad (now, Saint Petersburg). Its title is "Seated Dancer Adjusting Her Shoes." The resemblance was so close as if Degas had made the drawing to illustrate this chemical phenomenon. This gave me an idea.

I thought my observation would be helpful in teaching rotational isomerism and decided to publish it in the principal periodical of teaching chemistry, the American *Journal of Chemical Education*. I could not pay for the reproduction rights of the Degas images, so I asked my artist friend, Ferenc Lantos, to prepare sketches to convey the essence of the drawings. This is how my paper appeared in 1983. The journal put one of the Lantos sketches onto the cover of the issue. Soon another magazine, *Courier* of UNESCO reported about my observation and it reproduced Degas' originals. *Courier* appeared in over one hundred languages.

* * *

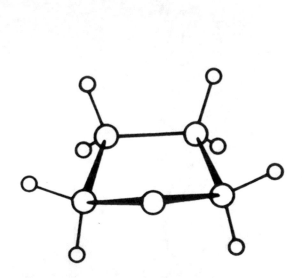

Model of the cyclopentane molecule and Henri Matisse's dancers at the Museum of Modern Art in New York (photograph by the author).

B–C bonds represent the legs. In the left hand drawing (previous page), the arms and the legs appear in a staggered way and in the right hand drawing in an eclipsed way. The two versions can transform into each other by rotating the arms and legs relative to each other. This is rotational isomerism, an essential phenomenon in chemistry, because one or the other of the same molecule represents differences in some of the chemical properties. It is so much easier to grasp the concept of rotational isomers through Degas' dancers than remaining in the realm of lifeless graphical representation. In addition to the Degas in Paris illustrating the staggered

Five carbon atoms form a ring in the cyclopentane molecule. Each carbon is linked to two hydrogen atoms, but the hydrogens are ignored in what I am going to discuss here. There are only four of the five carbons in the same plane at any given moment with the fifth out of the plane. In the next moment, the next carbon is out of the plane while there are still four carbons in the plane, and this keeps repeating from carbon to carbon. It would seem as if the five-membered carbon ring was rotating, whereas it is not. So it is not true rotation, only it gives the perception of rotation and it is called *pseudo rotation*. If there are five dancers in a circle and one jumps up and by the time this dancer returns to the floor, the next jumps, and this keeps repeating, we have a live model of pseudo rotation. This is what Matisse's dancers represent to me.

* * *

István Hargittai/Magdolna Hargittai

Symmetry through the Eyes of a Chemist

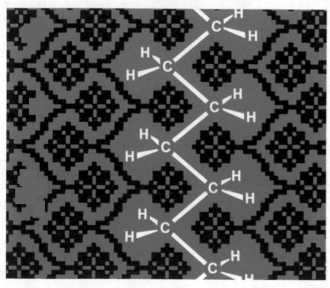

Cover of the first edition of our book *Symmetry through the Eyes of a Chemist* illustrating repetition covering the plane.

Another analogy can be applied to visualize crystal structures. They are characterized by repeating motifs in three directions, which is difficult to demonstrate, but it is easy to demonstrate such a repetition in one and two directions. There are altogether 7 possibilities to accomplish such repetition in one direction, and 17 possibilities in two directions. In the latter case the whole surface is covered by the repeating motif without gaps or overlaps. This kind of repetition creates a symmetric pattern and in our thoughts we can extend the operation into the third direction to produce the analogy with crystal structures. Already the one- and two-directional repetition provides a good perception of such infinite structures. When we subject a motif borrowed from Hungarian folk art to repetition, we arrive at attractive analogies for understanding the concept. This is what the renowned folk artist Györgyi Lengyel and I did, first for the repetition in one direction and then in two directions. We published our respective papers again in the *Journal of Chemical Education* in 1984 and 1985. The editors liked our compilations so much that they prepared one of the patterns and used it for cover illustration. A Hungarian folk art motif repeating in two directions appeared on the cover of the first edition of our book *Symmetry through the Eyes of a Chemist* published in 1986.[1]

---

[1] Third edition: Magdolna Hargittai and Istvan Hargittai, *Symmetry through the Eyes of a Chemist* (Springer 2009 for hard cover and 2010 for soft cover).

Lars Ernster in 1997 at the University of Stockholm (photograph by the author) and Raoul Wallenberg on November 26, 1944 (photograph by Tamás Veres, courtesy of the late Lars Ernster).

We met the famous Hungarian-Swedish biochemist Lars Ernster (1920–1998) and his violinist wife, Edit Wohl, on February 26, 1996, in Los Angeles, at the University of Southern California (USC). It was the day of my George A. Olah Lecture, the inaugural event of the annual George A. Olah Lectures established by USC to honor its first Nobel laureate (see the Olah chapter). Olah found it important that Magdi and I and the Ernsters met. It was the beginning of an intensive, albeit short, interaction.

When in spring 1996, the Ernsters returned to Stockholm, Lars talked with Torvard Laurent about me. Laurent was the President of the Royal Swedish Academy of Sciences at the time and he was also in charge of the Wenner-Gren Foundation. He was looking for a potential lecturer for the annual lectureship of the Foundation. Soon, he came to Budapest and we met. We both liked walking and a long walk gave us an opportunity to talk and talk. Following Torvard's return to

Stockholm, I received an invitation for the annual Wenner-Gren lectureship. The presentation was to be given in October 1996 at the Royal Swedish Academy of Sciences. The event took place in the evening of the day of the announcement of the 1996 Nobel Prizes of Physics and Chemistry, and the general topic of the presentation was to be symmetry. I could not know the areas in which the Nobel awards would be announced, but as it turned out, both the physics and the chemistry prizes were for symmetry-related discoveries. In 5 years, I was asked to give a yet more prestigious lecture in Stockholm (more about it in the Torvard Laurent chapter).

Lars Ernster was born in Budapest as László Ernster. He was 16 years old when his physician father died at the age of 48. After graduation from high school, Ernster would have liked to attend medical school, but the existing law of *numerus clausus* (closed number) made this impossible.

I. Hargittai, *Mosaic of a Scientific Life*, https://doi.org/10.1007/978-3-030-34766-6_7

This was a 1920 legislation, severely limiting the number of Jewish students that could enroll in higher education in Hungary. By now, it was 1938, and the *numerus clausus* was fast approaching *numerus nullus*, especially in medical schools. Ernster went to Paris and enrolled at the Sorbonne. To support himself, he was selling watches door to door in the evenings. When World War II broke out in 1939, he had to leave France, and he returned to Budapest. He worked as a laboratory assistant in the Jewish Hospital. He left, when he was called for slave labor, and then he was back at the Jewish Hospital until 1943. In the laboratory, he and another young man were assisting a physician who taught his helpers everything there was to know. It was good foresight, because the doctor was also called for slave labor and all the duties of the laboratory fell onto the two assistants. Ernster was not paid for this work, so he tutored pupils in the evenings. In 1944, the Arrow Cross—the Hungarian Nazis—murdered everybody, patients and personnel, in the Jewish Hospital. By then, Ernster was no longer there; he had found refuge in the Swedish Embassy.

Ernster's father-in-law, Hugo Wohl, was the managing director of the electronics company Orion, which held close business ties with Sweden. The Wohl family moved into the Embassy in June 1944. Raoul Wallenberg arrived in Budapest in July 1944 as a member of the Embassy staff and began his legendary humanitarian activities of saving Jews. Hugo Wohl became one of Wallenberg's closest associates during the following 6 months. On at least one occasion, Wallenberg personally extracted Ernster from the hands of the Arrow Cross. Even the Embassy was not considered safe enough during the last weeks of the Arrow Cross reign and the Wohl family, together with Wallenberg, moved to the cellar of a bank that was under Swedish protection. With the entry of the Red Army into Budapest, Wallenberg left from here on January 17, 1945, to meet with the Soviets. He vanished and was never heard from again. He fell victim to Stalin's tyranny.

Ernster and his wife left Hungary in April 1946. He first worked as a technician in Sweden until 1948 when his dream came true and enrolled at Stockholm University. At the time, there were restrictions in accepting foreigners at the medical school and he majored in chemistry. Later, he became an internationally renowned biochemist and was the Professor of Biochemistry at Stockholm University for two decades. His field of interest was bioenergetics and his discoveries found their way into textbooks. He was a long-time member of the Nobel Committee of Chemistry and was instrumental in recognizing discoveries in biochemistry with Nobel Prizes. He was one of the most decorated scientists in Sweden having received both Swedish and international prizes and awards. He had a large collection of the miniature Nobel Prize medals that the members of the Royal Swedish Academy of Sciences receive for their participation in Nobel decisions.

During our brief period of interactions, Ernster introduced me to many leading representatives of Swedish scientific life. His funeral gave me the opportunity to get to know the Kleins, George and Eva (see a separate chapter). It was as if Ernster wanted to make one more introduction for me. The crowd of international scientific luminaries who gathered to bid him farewell showed how much he was appreciated among his peers worldwide.

László Fejes Tóth

8

In Memoriam: The Lázár Family

László Fejes Tóth in 1999 in his home in Budapest (photograph by the author).

László Fejes Tóth (1915–2005) wrote a beautiful chapter for my first edited symmetry volume published in 1986 by Pergamon Press. I became first acquainted with his achievements by reading the famous Canadian Donald Coxeter's classic book on geometry. Coxeter quoted Fejes Tóth's statement about the human ability to create an infinite number of new (mathematical) worlds whose laws we may know but which we are unable to enter. I visited Fejes Tóth in 1999 and recorded a conversation with him.

He was born in Szeged and when he was 5 years old the family moved to Budapest. His father worked for the Hungarian Railways and acquired a law degree when he was 50 years old. Fejes Tóth's mother taught Hungarian and German in a high school for girls. Already in high school he became versed in calculus, which captivated him.

At the university, the renowned mathematician Leopold Fejér (born Weisz) was his mentor. It was not so much concrete mathematics, but the ability to become fascinated by the beauty of new discoveries is what he learned from Fejér. As a freshman, Fejes Tóth took up a problem and solved it. Here, I mention only the problem to indicate its flavor. Let us have an iron sphere whose temperature solely depends on the distance r measured from the center of the sphere. Immerse this iron sphere into water of 0 °C (32 °F) and the question is the temperature at the distance r after a given period of time. The problem was solved by the great Joseph Fourier at the end of the nineteenth century. Another great mathematician, Augustin-Louis Cauchy, found a solution more general than Fourier's. Fejes Tóth found a yet more general solution.

It was characteristic of Fejes Tóth's entire career that he posed challenging problems. However, the problem that stayed with him like a leitmotif originated from another mathematician, Dezső Lázár (1913–1943). Lázár posed the question about the way of positioning n points in a square or in a circle in such a way that the minimum distance between them would be maximal. The solution depended much on the shape of the surface in which the points would be positioned; hence a *general* solution could only be approximate. Exact solution could be found only for special cases. However, if n is sufficiently large, the problem can be formulated as follows: How to position circles of equal radius on the surface in the densest possible way? Fejes Tóth was the first who solved this problem, or so he thought. It turned out that a

8

© Springer Nature Switzerland AG 2020
I. Hargittai, *Mosaic of a Scientific Life*, https://doi.org/10.1007/978-3-030-34766-6_8

25

Norwegian mathematician, Axel Thue (1863–1922), had already solved it early in the twentieth century. This could have disappointed Fejes Tóth, but it did not; rather, he felt it as an inspiration for the continuation of this kind of research.

What Fejes Tóth told me about the fate of Dezső Lázár and his family would have been heartbreaking for anybody. It was so especially for me, because it was so close to what happened to us. Fejes Tóth and Lázár had known each other for a long time and they met again at the time when World War II had just begun. It was in Kolozsvár (Cluj in Romanian), a central city in Transylvania. It used to be part of the Hungarian Kingdom, then, following World War I, it was part of Romania until a Hitler-dominated treaty returned it to Hungary. The mathematician Lázár could not find employment in Hungary proper and landed a job in the Jewish high school in Kolozsvár. In 1942, he was called for slave labor and sent to the Russian front. He was ordered to sweep minefields with his bare hands. A mine exploded beneath him and lacking medical attention he bled to death. In 1944, his wife and two children were put into a cattle carriage and sent to Auschwitz where they were gassed upon arrival.

The story of the Lázár family resembles ours. My father, Dr. Jenő Wilhelm,[1] a Budapest lawyer, was called for slave labor and in September 1942 he was ordered to the Russian front to sweep a minefield with his bare hands. A mine exploded beneath him and lacking medical attention he bled to death. In June 1944, my mother, my 10-year old brother, and I (not quite yet 3) were put into a cattle carriage and sent to Auschwitz. It is at this point where our fates diverged from the Lázárs'. As we learned years later, the Hungarian Jewish leaders and the Germans negotiated for a few trainloads of Jews to be sent to labor camps in Austria. One of the selected trains had been sent to Auschwitz by error. A replacement train was needed and it was ours. My mother remembered only that at some point in our "journey" the train stopped, moved back some distance and started again. This must have been when it changed direction.

Returning to Dezső Lázár, he had only a single mathematical paper in 1936. It was in set theory. When Paul Erdős[2]

Dezső Lázár (source: http://www. komal.hu).

saw the paper in manuscript, he recognized its value, and showed it to John von Neumann[3] who arranged for its publication in *Compositio Mathematica*. Lázár had one more publication, posthumous, in 1947, organized by his friends. There are few memorials in public places remembering the hundreds of thousands of victims of the Hungarian Holocaust and yet fewer, remembering individuals. There is a memorial plaque at the Rényi Institute of Mathematics in Budapest with two sets of names of martyrs. One is for "Our Greats" and the other for those who had just "Embarked on the Road of Creating." Dezső Lázár's name is carved into this second list.

Fejes Tóth was proud of having been a Fejér pupil. Professor Fejér did not have many disciples and he followed their careers closely. Almost all became professors of mathematics in international universities. Fejes Tóth was an exception. He worked at the Rényi Institute and was its director for a while. He received the highest awards a scientist could have in Hungary. He considered his most valuable distinction though the invitation from the University of Zurich, which offered him a tenured professorship. He wanted to accept it, but he was not allowed to do so by the Hungarian authorities. He stayed in Budapest and thus he became an exception among Leopold Fejér's former pupils.

---

[1] My surname used to be Wilhelm until I changed it when I was 18.

[2] Paul Erdős (1913–1996) world-renowned mathematician; his mother, Anna Wilhelm, and my father, Jenő Wilhelm, were cousins.

[3] John von Neumann (1903–1957) was a Hungarian-American mathematician and computer scientist.

Árpád Furka

From Accelerated Matriculation to a Milestone Discovery

Árpád Furka in 1999 at Eötvös University, Budapest (photograph by the author).

Árpád Furka (1931–) is one of the most original scientists I have encountered. We first met in 1967 on the eve of Otto Bastiansen's visit (see a separate chapter). Furka had a new graphical method for determining structural parameters, such as the length of carbon–carbon chemical bonds. He thought this technique could replace the labor-intensive experimental and computational procedures. This was not feasible, but the idea was highly original. We have become friends. In the 1980s, he came up with a new technology for producing peptides and proteins, and he proved its feasibility in a meticulous way. Most of his immediate colleagues showed no interest or doubted what he said, but I knew he deserved all attention, because we might witness an extraordinary discovery.

Peptides consist of short chains of amino acids and proteins consist of long chains, and there is no sharp division between them. There are altogether 20 naturally occurring amino acids, so even small peptide molecules may exist in a great variety of their composition. There are two questions to ask. Which amino acids form the molecule that we happen to be interested in and what is the order in which they follow each other? If we want to produce even a small peptide in all its possible variations, one by one, it would take an enormous amount of time and labor. Furka estimated that to produce all variations of peptides consisting of five amino acids, would take 43,800 years for a well-trained chemist. Imagine how long it would take to produce larger peptides, let alone proteins, in all their variations.

Furka was fascinated by this question and realized that an entirely new approach was necessary to tackle such a task. He was thinking about it for years. In his thought experiments, he was producing the peptides of the same length but in a variety of composition, starting them all with one amino acid, then linking another one to each, and continuing the addition of one amino acid at a time. He was growing his peptides simultaneously. When he reached the desired chain length, he separated the chains. All the chains had the same composition, but they differed in the order of the amino acids. For the separation, he had worked out the necessary techniques as well. This approach has become known as combinatorial chemistry.

He became a discoverer against all odds. The conditions of his childhood and his career were full of barriers. The famous Hungarian expatriate scientists are often asked this question when they come home for a visit: "Could you have accomplished all what you did had you stayed in Hungary?"

© Springer Nature Switzerland AG 2020
I. Hargittai, *Mosaic of a Scientific Life*, https://doi.org/10.1007/978-3-030-34766-6_9

An honest response can only be a "No." However, one can never be sure in the absence of control experiments. To some extent, Furka's story may be considered to be such a control experiment.

He was born to Hungarian parents in Kristyor, Romania, was one of six siblings, and they lived in poverty. His father worked in a gold mine. His mother was originally from Hungary. The mother and five of the children returned to Hungary in 1942; the father and a married daughter remained in Romania. Furka completed his schooling in Hungary when he was 14 years old. He worked from early childhood helping the family to make ends meet, but he always regretted having had no opportunity to study, and he read everything that came his way. By 1950, he was already years behind his more fortunate peers. However, by then the communist regime decided to train members of the previously disadvantaged classes and create a new intellectual class. Furka was accepted to a school of accelerated education. They learned in 1 year what was prescribed for 4 years in the regular high school. The instructions were of high level, but they lacked all subjects that were considered expendable for a professional training. Not everybody could keep up with the elevated demands of such schools, but those who had the willpower and stamina, succeeded. Furka was among them.

After graduation, he was directed to continue at the University of Szeged and he received his Diploma (a master's degree equivalent) as a teacher of chemistry and physics. First, he taught in a high school of a provincial town. Soon, however, he was appointed to be an instructor at the University of Szeged, and from 1961, he worked at the Department of Organic Chemistry of Eötvös University in Budapest. He felt alien among his colleagues. He realized that he missed those "superfluous" subjects that were absent from his high school curriculum. His situation may be characterized with the following episode. The professor in charge of the Department was a scientist of the old class, Győző Bruckner (1900–1980). The Ministry of Education forced Furka down his throat—Bruckner had considered someone else for the position. Incidentally, as a sophomore, I attended Bruckner's superb lectures in organic chemistry. Bruckner used to have the leading associates of his department for a weekly tea and, overcoming his resentment, he sent word to Furka that he invited him for tea. Furka replied that he did not drink tea. This was true, and Furka realized later that this was a social occasion and drinking tea was its least important part.

The recognition of Furka's discovery resulting in combinatorial chemistry was slow, especially in Hungary, and he was never elected a member of the Hungarian Academy of Sciences. Unfortunate circumstances hindered proper evaluation of his contributions. That included publications by others that Furka characterized as plagiarism of his work. It was only thanks to the honesty of the international scientific community that his achievement and his name did not disappear into oblivion. By now, there are references to him as the "father" or "pioneer" of combinatorial chemistry. I have helped with two papers to set the record straight stressing that it all started with his ideas and publications. Although he has been in retirement for many years, his activities have not diminished. He publishes papers of original research, and has produced a book about the Universe, a most readable account of science history. It would still be possible to award him the Nobel Prize, and it would be well deserved. I am not making such a suggestion lightly, and there is no similar suggestion elsewhere in this volume.

Pál Gadó

He Taught us Democracy

A snapshot of Pál Gadó (courtesy
of the late Beáta Ignácz).

Pál Gadó (1933–2016) attended the famous Lutheran High
School in Budapest where luminaries of science had studied,
such as Eugene P. Wigner (see a separate chapter) and John
von Neumann. Then, Gadó studied physics at Eötvös University. His research was in materials science and technology
and his particular interest was bauxite, the raw material for
aluminum. He had a severe physical disability, but it did not
slow him down. When he was in his fifties, at the top of his
career in science, he changed course, and took up an active
function in the organization of people with disabilities.

Our paths crossed in the early 1970s when we both were
members of the scientific society for measuring devices and
automation. I remember in particular one meeting when our
discussion of a topic stretched on and we still could not come
to an agreement. Someone lamented that the chairman, Gadó,
could have already decided which way to go or we could
have decided the question with a majority vote. It was then
that Gadó told us that the democratic approach means longer
debates, and it is more painful than having the chairman
decide or even a majority vote. If we are aiming at democracy
(I do not think the official line was that we were, then), it
cannot be the dictatorship of one, not even the dictatorship of
the majority. We have to listen to the minority opinions. The
question under discussion was not at all of real importance,

but the lesson the exchange offered was. For me, it was a life
lesson. Soon enough I could test whether I could apply it.

There was a committee meeting within the organization of
the international crystallographic community. We were
immersed in a long discussion over a not too significant
issue. The Swiss delegate proposed to end the endless discussion and decide the issue by vote. He would accept the
outcome whether his opinion prevailed or not, because he
was used to democracy where all fall in line dictated by the
majority decisions. At that point, I reiterated what I had
learned from Gadó, and our Swiss colleague agreed that we
might continue our discussion a little longer.

Remembering Gadó, I could not be the only one whom he
made to think about the meaning of democracy. My generation grew up under the conditions of a totalitarian regime in
Hungary, and what was natural for people who grew up in a
democracy, was not necessarily obvious to us even though
most of us were thriving toward a democratic ideal.

There was a Moscow chemist, Evgeny Shustorovich,
whose books I liked a great deal and not only for their
chemistry. In one of his books, every chapter opened with a
statement of emphasis. One of them caught my imagination
especially. It was about the decisive influence of the
circumstances under which one begins a career in science.

At that time I could quote it verbatim. Of course, it did not say anything extraordinary, but somehow it resonated with me. I had never met Shustorovich and then learned that he and his family left the Soviet Union sometime in the early 1980s and settled in the USA. Announcing one's intention to emigrate was a big deal in the Soviet Union. It could be followed by years without much of what people usually take for granted under normal circumstances. This means that those who declared their intention to leave must have been very determined to change their lives. When they were finally allowed to go, they had restrictions in what they could take with them. This is what gave me the idea that Shustorovich may have not taken his books with him as he may have given other things higher priority in starting a new life. So I decided that the next time we would go to the USA, I would take with me my Shustorovich books and present them to him.

We visited the Shustorovich family on the next occasion. They lived in Rochester, New York. He was rather taken aback by my gift of his books. He asked me why I singled out that particular chapter opening. I told him that if we were aware of the limitations of our beginnings, it may be easier to break away from those limitations if we wanted to break away from them in the first place. I certainly wanted to break away and was sure that so did he.

Shustorovich had a good job; he was doing theoretical research in coordination chemistry at Kodak. I would have loved to see him in a university setting, but, perhaps, his English was not up to it. To me, a university professorship provides the highest degree of academic independence, which I value greatly. Soon I understood that Shustorovich could also use a bit more academic freedom. In 1985, I was organizing a big volume on symmetry, collecting contributions from the most diverse authors and disciplines. Shustorovich was among the first batch of potential authors I invited to participate. He did not respond promptly and when he did, it was a disappointment. He had asked Kodak about his participation in the symmetry project and Kodak did not see any advantage for the company in his participation. Thus, Sustorovich declined, very regretfully. It must have been a bitter experience for him to receive the company response after which he did not find it proper to contribute even as a private scholar. Following his immigration, as far as I know, he hardly published anything beyond his narrow specialty in science. It would be a sad conclusion that the adverse conditions in the Soviet Union might have enhanced his literary production.

Nothing in my experience has diminished the validity of Shutorovich's maxim about the importance of the circumstances of the beginnings of a research career. Discussions, debates, constructive criticism, the possibility of testing ideas, hypotheses, models, and theories are all among the necessary ingredients for creative activities. Lacking them may be an impediment to advancement. In 1998, I visited James W. Black (1924–2010) in his London laboratory. He received a share of the Nobel Prize in Physiology or Medicine 10 years before. Gertrude B. Elion and George H. Hitchings were his co-laureates. The motivation for the award was their discoveries of principles for drug treatment and all three had discovered specific new medicines. Black is most famous for the beta receptor blocking propranolol (for many years, I have been taking a pill of it every morning) and for cimetidine for treating peptic ulcer. Black did not have support for his university studies and accumulated a large debt by the time of his graduation. Although he was most interested in research in physiology, he accepted a teaching position at the medical school in Singapore. There, he continued his research. He designed a new device for measuring blood pressure and investigating blood circulation in animal experiments. He amassed a great amount of data. Alas, he did not quite know what to do with them. There was no one with whom he could have discussed his work, no one who could have questioned the conclusions he was drawing from his observations. He was bouncing between two extremes, the euphoria of everything is wonderful and the despair of nothing is of any value. Luckily, he did not throw away his data and upon his return to Great Britain, he continued his work embedded in a research environment.

Scientific discussions have great value if indeed only considerations of science are involved in them. Even democratic considerations have little role in scientific debates. Nothing can be decided by vote. Authority has or should have no role in them either. A good scientist would never try to force his or her opinion onto a disciple. Nonetheless, it happens that the elder scientist has such great authority that even unwittingly such an influence impacts the development of a project. This happened when the young Eugene Wigner had some innovative ideas, but when he sensed his mentor's, Michael Polanyi's, indifference, Wigner let the idea drop. Eventually, it turned out that the idea should have been vigorously pursued. Something similar happened between Lev Landau and his young associate Aleksei Abrikosov, both physicists and both future Nobel laureates. In both stories, it was not the explicit opinion but the authority of the mentor that influenced the younger scientist's decision not to continue something that eventually turned out to be of seminal importance.

The Royal Princess Chulabhorn of Thailand and the author in 1999 at the Chulabhorn Research Institute in Bangkok (photograph by Magdolna Hargittai).

We observed an entirely different manifestation of the unnatural presence of authority at the Chulabhorn Research Institute in Bangkok, Thailand. The relationship of the head of the Institute, the Royal Princess Chulabhorn (1957–), a PhD chemist, with the rest of the institute is determined by obsolete rules governing the interaction between the members of the royal family and common people. The deputy director of the institute, herself a notable scientist, approached the princess crawling on her knees every time they had to discuss something. Under such circumstance, no real scientific discussion could be imagined. Chulabhorn is a bona fide researcher. She earned her PhD degree in Thailand. She did postdoctoral studies at the University of Ulm in Germany and she had a study visit to Tokyo where she carried out research at the medical school. Her principal research interest is in the chemistry of biologically active natural products. Thailand has a rich indigenous fauna and flora.

I conclude with a strange experience of censorship at the now defunct magazine *Chemical Heritage*. This was something I, behind the "Iron Curtain," could have been reluctant to imagine to happen in the "free world." I recorded a conversation with Koji Nakanishi (1925–2019) in 2002. He immigrated to the USA from Japan in 1969 and became a professor of organic chemistry at Columbia University. In our conversation we talked about Japan's role in World War II, and according to Nakanishi, prior to the war, "The Americans were overwhelming Japan with their power." "But did not Japan start the war against America?"—I interjected. To which he responded: "That is so, but the Japanese were slowly being strangled by the petroleum policy of the United States." The magazine *Chemical Heritage* wanted me to remove this part from the interview before publication. They told me that the magazine could not be identified with such views. However, it was not the magazine's view, it was Nakanishi's. Nonetheless, the magazine did not want to have this on its pages. To me, on the contrary, it was of interest to see that even after having lived in New York for over three decades, this renowned scientist was still advocating the Japanese excuse for attacking America, not to speak about how it was actually done. I thought it was of interest to Nakanishi's long-time colleagues as well as his many former students to learn about his views. I may add that this was not something uttered on the spur of the moment. Rather, as I always did, after having transcribed the conversation, I sent the text for inspection to Nakanishi.

He could change, modify, replace whatever there was in the transcripts, but he did not alter anything of significance. Finally, *Chemical Heritage* turned to Alfred Bader, the sponsor of the column, "Hargittai Interviews." Bader's opinion was that deleting the section from the interview would be tantamount to its falsification. Nonetheless, the magazine printed the censored version of the conversation.[1]

---

[1] I. Hargittai, "Koji Nakanishi," *Chemical Heritage* 2003, 21(3), 6–9. I thank Ashley Augustyniak of the Science History Institute, Philadelphia, for a copy of the publication. *Chemical Heritage* used to be the magazine of the Chemical Heritage Foundation—now the Science History Institute.

They Could Have Been Hungarian Scientists

In Fort Lee, New Jersey, 2010, from left to right, Endre Szemerédi, Lori Lax, Peter Lax, Richard Garwin, Anna Kepes (Szemerédi's wife), Magdolna Hargittai, Lois Garwin, and István Hargittai (photograph by Eszter Hargittai).

Enrico Fermi considered Richard L. Garwin (1928–) the only genius whom Fermi ever met. This statement cannot be taken lightly because Fermi was not known for exaggerations; and he had met many outstanding scientists. Garwin worked with Fermi during the post-war years. We first visited the Garwins in their home in Scarsdale, New York, in August 2004. From that time, for the next decade, we met with them annually.

Garwin's best-known work was when he, as a young scientist, designed the technical blueprint of the world's first hydrogen bomb in the Los Alamos nuclear laboratory—today it is the Los Alamos National Laboratory. The principle of the hydrogen bomb was described by Edward Teller and Stanislaw Ulam in a joint report, which is still classified. The 24-year-old Garwin provided the technical solution, which has also remained classified. His name and his role in developing the hydrogen bomb have remained in the shadows, in part, because Teller almost monopolized the limelight and, in part, because Garwin himself was not eager to be known for this contribution. He opposed the deployment of the hydrogen bomb. During the next decades, Garwin continued his participation in the American weapons programs, but his principal employment was with IBM. When he was needed for the weapons program, IBM "lent" him for it. Garwin was known to have developed efficient

© Springer Nature Switzerland AG 2020

I. Hargittai, *Mosaic of a Scientific Life*, https://doi.org/10.1007/978-3-030-34766-6_11

interactions with other renowned physicists. According to one of them, Valentine Telegdi,[1] when Garwin participated in an experimental research project, he solved all the problems before the others might have had a chance to shine.

Garwin was a staunch critic of the so-called Star Wars concept—the Strategic Defense Initiative—and clashed with Edward Teller. They conducted some of their debates in the pages of *The New York Times*. Their styles as debaters differed sharply. Teller was not always factual and one of his favorite tools was intimation. Rather than providing evidence he would hint that he had it, but could not reveal it because the information was classified. It happened at least once that another well-known scientist walked off the stage of the debate because he found it impossible to conduct the debate fairly under such conditions. Garwin's approach was opposite to Teller's, factual to the limit, which made his reasoning less colorful. Teller was a legendary debater; according to some he was invincible. However, Garwin was not intimidated by Teller's fame.

Garwin has enjoyed a high reputation and respect by the American physicists, and also by American politicians and military leaders. Until quite recently, he served on committees of experts advising US presidents. Because of his background, I found his statement made during our conversation in 2004 frightening. He said "I do believe that we are lucky not to already have had a terrorist nuclear explosion in one of our cities. I confidently believe that we will have one within the next few years."[2] Ten years later I asked him about this, and he still thought there would be such an attack, but he was no longer so convinced in 2014 as he was in 2004.

On his maternal side, Garwin was of Hungarian origin. His mother, Leona Schwartz, was born in 1900 in Hungary and died in 1996 in the USA. Garwin learned only in 1985 that his mother was not born in America. Indeed, she was 12 years old when she immigrated together with her family. They spoke Hungarian at home. Her first job was in a retail store and her supervisor was happy with her. One day he told her that Monday mornings she hardly speaks English, but come Friday her language improves only to start everything again Monday morning. From that day, she never spoke a word in Hungarian.

There have been many Hungarian expatriate scientists who became successful in the USA and Western Europe. The exodus of Hungarian intellectuals continues to this day. These expatriates benefited a great deal from the education they received in Hungary, so the pride is justified when they are included in the enumeration of successful Hungarian scientists. The real question is in the emphasis and the motivation in blending the performance of expatriates with those of the scientists at home. If it is pride and appreciation, that is alright. If it is for creating a myth of cultural superiority, that is wrong. If the inclusion then goes without recognizing the reason for emigration—often, though not always, it has been anti-Semitism—then it is tantamount to falsifying history.

Here I introduce the reader to five scientists, all connected to Hungary one way or another, who had brilliant careers. I have had interactions with them.

The Jewish Hungarian family of the paternal grandfather of the biochemist Robert F. Furchgott (1916–2009) used to live in Nyitra (today, Nitra, in Slovakia) from where they immigrated to America. Part of the family lived in Budapest, including a professor of astronomy. The original surname was Fürchgott and there are Fürchgott graves in the Nitra Jewish cemetery. Furchgott spent most of his career at the Department of Pharmacology of State University of New York (SUNY). This department is on the Brooklyn

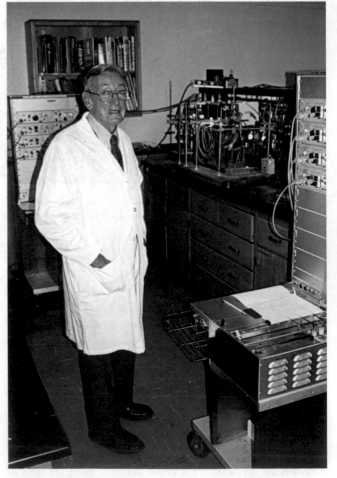

Robert F. Furchgott in 2000 in his laboratory in Brooklyn (photograph by the author).

---

[1] Valentine Telegdi (1922–2006) was a Hungarian-born American physicist.

[2] I. Hargittai, M. Hargittai, *Candid Science VI: More conversations with famous scientists*. Chap. 23: Richard L. Garwin (London: Imperial College Press, 2006); pp. 480–517; actual quote, p. 516.

campus of SUNY. We visited Furchgott in 2000. His research area was the physiology of the cardiovascular system. He was the first who determined that there was a signal carrying substance that later turned out to be NO, the nitrogen oxide molecule. He and two other scientists shared the 1998 Nobel Prize in Physiology or Medicine. There have been debates about whether the other two had been the most deserving individuals to share the prize with Furchgott. Nobody has ever questioned that he was the principal player in this story. Furchgott, as many others, opined that Salvador Moncada (1944–) should have been one of the awardees. Moncada is a British scientist from Central America.

Laurie M. Brown (1923–2019) was born in New York City. His paternal grandfather studied in what is today Bratislava, the capital of Slovakia (then, Pozsony, part of Hungary) to become a cantor. They spoke Yiddish and Hungarian at home. Brown's parents were both from Hungary,

Laurie M. Brown in 2003 (courtesy of Laurie M. Brown).

but met in America. They spoke Hungarian; especially when they did not want their children to understand what they were talking about. They could not afford college, but made sure their children got a higher education. They lived in Brooklyn and Brown attended the Manual Training High School between 1936 and 1940 where the emphasis was on learning practical crafts. It was the same school that the world-renowned physicist Isidor I. Rabi attended a generation before. It might have not been a strong school academically except that at Brown's time—the time of deep economic crisis—highly qualified teachers found jobs in this school. Brown majored in physics at Cornell University and did his doctoral work under Richard Feynman. Brown spent his entire career at Northwestern University in Evanston, Illinois. He was a nuclear physicist and did research in elementary particles as well. He had visiting stints at renowned institutions, such as the Princeton Institute for Advanced Study, the University of Vienna, and the Argonne National Laboratory. Lately, his interests had shifted toward the

history of quantum physics and elementary particles and he became an authority in science history. In addition to Feynman, he published jointly with other noted physicists, among them, the Nobel laureate Yoichiro Nambu.

Charles Weissmann (1931–), a world-renowned Swiss molecular biologist and pharmacologist, is often counted among the Hungarian-born scientists. His parents were Swiss and when his mother was expecting, she visited a lady friend in Budapest. That is when Charles was born.

Charles Weissmann in 2000 in his office in London (photograph by the author).

After the birth she returned to Switzerland. He studied at the University of Zurich, earned an MD degree and soon after, in 1961, a PhD in chemistry under the mentorship of the Nobel laureate Paul Karrer. For 6 years Weissmann did postdoctoral studies with the Spanish-American Nobel laureate Severo Ochoa at New York University (NYU). Upon Weissmann's return to Switzerland, he was appointed to head the Institute of Molecular Biology of the University of Zurich. His most conspicuous discoveries came about in molecular genetics and in neurodegenerative diseases. When he reached the Swiss retirement age, he moved to London and for years he was in charge of the laboratory of neurogenetics of the British Medical Research Council. When he reached the British retirement age, he moved to the USA where there is no mandatory retirement age, and became laboratory director at Scripps Florida. Only in 2017 did he become Professor Emeritus, at the Department of Immunology and Microbiology. There is another Hungarian relevance in Weissmann's life in addition to his place of birth. His present (second) wife escaped deportation to Auschwitz in 1944 as a result of the negotiations between the Hungarian Jewish leaders and the Nazis. The agreement was to exchange a trainload of Jews for goods.

Gabor Somorjai (1935–) is professor of chemistry emeritus at the University of California at Berkeley. He is a world

Gabor Somorjai in 2000 at the Budapest University of Technology and Economics (photograph by the author).

George Radda in 2000 in the office of the President of the British Medical Research Council in London (photograph by the author).

authority in researching the chemical reactions on solid surfaces. He has excelled both in experimental and theoretical discoveries and in their practical applications. He was born in Budapest and his life and the life of his immediate family members were saved by Raoul Wallenberg (see more on Wallenberg in the Ernster chapter). Many of his relatives perished in Auschwitz. Under the communist regime in Hungary, he was first refused acceptance for university studies. However, his father bribed some influential people and Gabor could enroll at the Budapest Technical University (as it was called then). Following the suppression of the Revolution in 1956, he moved to the USA as a refugee and graduated from Berkeley. He had a brilliant career, was elected a member of the National Academy of Sciences of the U.S.A. at the age of 44; and received the National Medal of Science from President George W. Bush in 2002. He could have shared the 2007 Nobel Prize in Chemistry with Gerhard Ertl (1936–) who received it for his studies of chemical processes on solid surfaces. The scientific community of the field was taken by surprise that Somorjai was left out of this recognition, which he would have amply deserved, and there were protests.

George K. (György Károly) Radda (1936–) was born in the northwestern Hungarian town Győr. He was a student of Eötvös University, like Somorjai, he also became a refugee following the suppression of the 1956 Revolution. He completed his studies at Oxford University where he was appointed professor in 1984. First, he was a chemist, then a biochemist, and finally a physiologist whose research has focused on the heart using magnetic resonance imaging (MRI). His science did not fit any of the traditional disciplines and a new chair was created for him: he became the professor of molecular cardiology. He was the president of the British Medical Research Council between 1996 and 2004. This was the institution that initiated the Laboratory of Molecular Biology in Cambridge after the war in which by now a dozen scientists have been awarded Nobel Prizes. I visited Radda in 2000 in his office. He told me that he considered it the task of the Council to support proposals of uncertain outcome thereby encouraging researchers to take on innovative projects. Currently, as an emeritus, he is advising researchers, dividing his time between Oxford and Singapore.

Ronald J. Gillespie with the author at an international meeting in Austin, Texas (by unknown photographer).

My first trip to the USA in January 1969 was for a 1-year visiting position at the Department of Physics, University of Texas at Austin. On my way there, I stopped to give a talk at Indiana University in Bloomington. At that time visitors from Eastern Europe were a rarity. I talked about our experimental results and somebody asked whether I had tried to apply Gillespie's model for the interpretation of our data. No, I did not, but I stayed up all night and read everything about Gillespie's model in the library. It had a strange name,

valence shell electron pair repulsion (VSEPR) model. It was eminently applicable to our findings. That very night I composed a long letter to Magdi who was already my wife and my graduate student. I illustrated the letter with carefully colored sketches.

The geometry of a molecule—the spatial arrangement of its atoms—is responsible to a great extent for much of the chemical behavior of the molecules. For simple molecules, it is determined by the number of electron pairs in the

I. Hargittai, *Mosaic of a Scientific Life*, https://doi.org/10.1007/978-3-030-34766-6_12

outermost shell, the valence shell of the central atom. The VSEPR model makes it simple to predict the molecular geometry just by knowing how many electron pairs are there in the valence shell of the central atom. Electron pairs between two atoms form a bond, each contributing one electron. There are then the so-called lone pairs of electrons, of which both electrons belong to the same atom. Both the bonding pairs and the lone pairs must be considered when predicting molecular geometry. The reason is that both take up space in the valence shell. This was an essential feature of Gillespie's model. Depending on the total number of electron pairs, there are linear, triangular, tetrahedral, trigonal bipyramidal, and octahedral molecules when there are two to six electron pairs to be accommodated. We may produce the same arrangements when tying together two to six balloons of the fat, round type, because in this case the balloons, rather than the electron pairs, will be elbowing each other for space, but the principle will be the same.

When walnuts grow together, they take the same shapes as molecules or balloon clusters. This is my photograph, but the idea came from a Transylvanian chemist who first communicated photographs of walnut clusters as an analogy to molecular structures.

The shapes of molecules, balloon or walnut clusters are consistently the same, but they may not necessarily be what our intuition would expect. When, for example, tying four balloons together, we might expect the four balloons to form a square in which every balloon would take the position of one of the four corners of a square. However, it takes some force to make the balloons remain in such an arrangement. When the four balloons are left alone, they form a tetrahedron. Admittedly, the geometry of molecules depends on more factors than just the number of electron pairs. It is a simplification, but this is the essence of building models. Choose the most decisive factor and neglect the rest, and if such an approach yields good predictions, it is a workable model.

A British scientist, Ronald J. Gillespie (1924–), and an Australian scientist, Ronald S. Nyholm (1917–1971), described the VSEPR model in 1957 in a long article. Gillespie then further developed the model and published a book about it in 1972. Today, the model is discussed in every general chemistry text and is part of the curriculum both in high school and in college. When I returned from my 1 year in Texas to Budapest, I wanted to popularize the model and composed an article for the Hungarian popular science magazine called *Természet Világa* (World of Nature). Obviously, I wanted to demonstrate the utility of the model for the simplest possible systems. To my great surprise, the model did not work for some of the simplest molecules. I turned to Gillespie and he took my concerns seriously but asked me not to rush publicizing my observations because he did not want to hurt the book he had just published. I was not eager to publicize my observations before finding some explanation for the discrepancy. I did, however, publish my observations in the Hungarian magazine. Years later, having already found an explanation I could augment the tenets of the model that made its application more reliable. It was a recognition by Gillespie when he suggested that we co-author a new book about the model. It appeared in 1991 and Dover re-published it in 2012.[1]

Gillespie's model was a success story already at the early stages of its appearance. He could even suggest corrections to conclusions drawn from reliable experimental data. The experiments were trustworthy, but the interpretation was in error and Gillespie sent back the researchers to reinterpret their observations. Unfortunately, one of Gillespie's departmental colleagues kept annoying him by urging him to make the model more sophisticated. This colleague did not see that the beauty of the model was in its simplicity. Gillespie could not resist this bullying and was looking for a more sophisticated approach thereby losing some of the model's merits. This was not the only area in science where he did not find complete satisfaction. On one occasion, he poured out the origins of his frustration to me. The occasion was our recording a conversation for the *Candid Science* books series of interviews with famous scientists.[2] This was soon after George A. Olah received the Nobel Prize in Chemistry.

Gillespie found Olah's award well deserved but noted a marked difference between their approaches to research. They both had been engaged in working with superacids—acids stronger than the concentrated sulfuric acid. Olah used

[1] R. J. Gillespie and I. Hargittai, *The VSEPR Model of Molecular Geometry* (Allyn and Bacon, 1991; Dover, 2012).

[2] B. Hargittai, I. Hargittai, M. Hargittai, *Candid Science: Conversations with Famous Scientists*, Volumes I–VI (London: Imperial College Press, 2000–2006).

superacids for prolonging the lifetime of unstable reaction intermediates, thereby solving some previous controversies between famous scientists. Also, he developed a whole new chemistry from such research (see the Olah chapter). During the time when they both worked in Canada, Gillespie provided crucial assistance to Olah, letting him use expensive instrumentation that Gillespie had and Olah did not. Gillespie noticed that Olah's samples were not always purified as rigorously as Gillespie would have insisted on. As it turned out, Olah identified in these samples the intermediates of milestone importance that Gillespie might have "purified out" had he worked on those samples. This thriving for "perfection," whether it was his simple model or his chemical samples, constituted limitations in his achievements. Gillespie seemed to realize this "shortcoming" yet could not help following his restrictive approach whereas Olah let his imagination fly more freely.

There Was Once a Diplomat ...

André Goodfriend in November 2017 in Washington (photograph by Magdolna Hargittai).

André Goodfriend was well known in Hungary a few years ago. For many, he was a popular hero; others regarded him as the enemy number one. The USA did not have an ambassador in Budapest for a long time. The Embassy was led by André Goodfriend, the chargé d'affaires, for a year and a half in 2013–2015. When he was originally appointed to his post in Budapest, he anticipated a different task. He had been a diplomat for a long time, but he specialized in consular matters. He had expected to be charged with running the Embassy rather than fulfilling a more visible function.

Goodfriend was born in Los Angeles. His father immigrated to America as a member of a European refugee family after World War II. His mother was American born. André grew up as a true American youth with a diverse European cultural background. His studies included the classics, Greek and French, philosophy, and radio and television. He obtained his first degree in telecommunications. Then, he earned a master's degree in the sociology of telecommunications. He worked in the field of radio and television, as well as in higher education. He was already in his doctoral studies when the offer came and he moved into the diplomatic service. Prior to Budapest, his stations included Israel, India, Moscow, London, Syria, and he had other assignments as well. His Budapest posting was a promotion.

In the absence of an ambassador, especially when it became clear that there were no plans to fill the position, André's role changed. It pleased him to have received a task in real diplomacy. While he was in charge of the US Embassy, we met twice at some events in Budapest and each time we had a chat. One was when in January 2014 I gave a talk on the comparison of the American and Soviet nuclear laboratories, Los Alamos and Arzamas-16, at the Central European University (CEU). The President of CEU asked me to give this talk in connection with the publication of my book on Soviet scientists, *Buried Glory*. There is more about this book in the Andrei Sakharov chapter. André was sitting in the first row and the President of CEU introduced us to each other. At the end of the usual superficial conversation we agreed to come together for coffee, but it did not happen. The other time, a bit later, was a jubilee dinner for a Hungarian–American scholarship program. I was previously a member of the board of the organization and so I received an invitation. Goodfriend spoke at the meeting, and this was yet at the beginning of his political appearances. He expressed his criticism in the direction of the Hungarian government, but he spoke cautiously. One had to be able to read between the lines to understand what was being said. This time we talked longer and I checked with him whether I

I. Hargittai, *Mosaic of a Scientific Life*, https://doi.org/10.1007/978-3-030-34766-6_13

understood properly what he really meant to say. It appeared that I did. Again, we agreed that we should talk more over coffee, but, again, nothing came of it. Soon we left for the USA for several months. When we returned to Budapest, André was already packing or may have already left.

I learned that during our absence André had become a celebrity in Hungary. What I heard about his political activities was consistent with his speech at our second meeting, but there was no longer need to read between the lines. Of course, I knew about some of it as I was following the events in Hungary from America, however superficially. Thus, for example, I read about the statement by the American Embassy protesting the commemoration of the Nazi ally Nicholas Horthy with a bust in Budapest on November 3, 2014. Nicholas Horthy was Regent (head of state) of the Kingdom of Hungary in the period 1920–1944. The Americans called for a decisive and unambiguous condemnation of the commemoration by the highest-ranking Hungarian leaders. That did not happen. In 2017, the Prime Minister stirred sentiments by complimenting Nicholas Horthy.

André's most memorable moment came when the USA announced that six Hungarian officials would not be allowed to enter the USA because of their involvement in corruption. The identities of the six corrupt officials, who were not among the top political leaders, were not disclosed. It turned out that one of them was the head of the State Tax Authority as she herself revealed being one of the six. She wanted to see the head of the US Mission, ostensibly, to clarify the situation. As she and her lawyer were walking across Szabadság (Liberty) Square toward the Embassy, they met with André who was taking a stroll in the beautiful weather. What happened was as if it had been choreographed for a burlesque. The tax chief stopped André and demanded an interpreter to communicate what she wanted to tell him and to be able to understand what André was to tell her. Someone filmed the encounter, which then became the most popular YouTube entry. Even those who had not been acquainted with André, learned about him. The tax chief managed to become the laughingstock for the larger part of the population. The media, including the meager opposition media, were busy wondering who the other five corrupt officials might be and, especially, what proofs the Americans might have against them. However, I thought that their interest was misdirected. I thought then, and think now, that the purpose of the American action was issuing a warning. If the unlawful activities of these middle-level officials are known to the Americans, what may they know about the possible corruption involvement of the highest leaders of the country?

André Goodfriend with the author in fall 2015, in Duncansville, Pennsylvania (photograph by Magdolna Hargittai).

I regretted that our coffee meeting with André did not happen. When we were in New York next time, I wrote to him that we were there. So, André came up from Washington and we had not just coffee, but a full meal together. Our conversation did not seem to come to an end but he was due back in Washington so we had to stop. We wanted it to continue and thought if we enjoyed it so much, others might as well. This is how the idea of a book of conversations was born. As André was still in diplomatic service he had to ask the State Department for permission. It was swiftly granted with the stipulation that before publication, the manuscript of the book must be submitted for review to the State Department.

In September 2015, on two occasions, we spent each time two full days recording our conversations. For 2 days we went to André's residence and for 2 days André came to our son Balázs's house in Pennsylvania. These visits were followed by intensive correspondence until the manuscript was completed. As it turned out, none of us had thought that such a comprehensive, meaningful, and normal-size book could come together from these conversations, but it did.

André submitted the polished English manuscript to the State Department. We wanted to base the Hungarian translation on the approved English version. We already had a publisher for it in Budapest. This time it took the State Department much longer to come to a decision, and it was not to publish the book. This option was implied in the original permission. Although I was disappointed, I also understood it. I realized that there could only be one voice in Budapest for American diplomacy, and that only voice should be the voice of the American Ambassador who, eventually, arrived in Budapest.

Still I felt that the project was an enriching adventure. I asked someone, provoked someone, and argued with someone who was as prepared as possible for this exchange and articulated his message brilliantly. He did not brush off any of my questions or comments, even if he did not answer all my questions as deeply as I wanted. He always talked to me and never at me. I concluded from these conversations that André should teach Open Government and Open Diplomacy at a university level. I have no idea whether such a course exists anywhere, but if it does not, it should.

During the past years, we have met several times in Budapest, New York, and Washington. On each occasion, I was impressed with the intensity of his interest in everything Hungarian. I have also experienced how vividly his memory lives on in Hungary. Last time we met in New York, it was at the Consulate General of Hungary where I gave a talk about our book with Magdi, *Budapest Scientific*. It was organized by the New York Hungarian Scientific Society. When André showed up, the Consul General joined the audience. He was surprised by André's presence, but André told him that we knew each other. I wonder if this encounter was the cause that when the possibility of another talk was raised by the New York Hungarian Scientific Society, this time about our new book, *New York Scientific*, the Consulate showed no interest.

As I am completing this chapter in November 2017, we just had our latest meeting with André a few weeks ago. He attended our talks at the American Physics Institute near Washington, DC. Under the changing conditions in the USA, André is still charged with working on the possibilities of modern diplomacy and is directing an office of e-diplomacy. I trust that his innovative ideas will eventually be turned into practice. Also, I think that when he will no longer be in the diplomatic service, and diplomats usually retire earlier than most others, we might publish our book. Its lessons will not become obsolete any time soon.

Following up Our Lives

Árpád Göncz on November 10, 2003, in the Hungarian Parliament during the First World Science Forum (photograph by the author).

Árpád Göncz (1922–2015), a politician of the liberal party, *Free Democrats*, was President of the Hungarian Republic between 1990 and 2000, the first president after the 1989/90 political changes. He was popular, especially in those circles that I found close to my values. He was not a bona fide politician, but he was a bona fide human being. When he saw or experienced something condemnable, he did not hesitate to condemn it. When I completed the manuscript of my semiautobiographical book, *Our Lives*, it occurred to me that somebody should preface it. I thought to ask Mr. Göncz, but I hesitated. I had never had any personal interaction with him. I knew that if he would not accept my invitation, I would not find anybody as appropriate as him. But he accepted.

The subtitle of the book is *Encounters of a Scientist*.[1] It consists of 19 chapters and each chapter is titled after a Nobel laureate scientist whom I knew and with whom I had had some interaction. Each chapter begins with some science-related topic and moves gradually to some aspects of my life or the lives of my friends and family members. There is a great deal in the book about my parents and the history of our family, reaching back not very far. I kept it to a level that I thought our grandchildren—who at the time of the writing did not yet exist—might be interested in. Politics and history encroached to such a great extent on our story that made the involvement of Árpád Göncz yet more attractive.

I talked about my plans for the book to James D. Watson when he and his wife visited us in Budapest in the summer of 2000. They invited us to Cold Spring Harbor to work on the book. This is how we spent 3 months there in 2002. My office was in the press office of the Cold Spring Harbor Laboratory. I had an assistant, and an editor helped me. I never had such ideal conditions in working on a book before (or since). I prepared the first version of the manuscript in English as I had in mind our would-be grandchildren as my primary audience and I knew they would hardly read books in Hungarian. Then it happened that the book first appeared in Hungarian translation in 2003, and a year later, in English.

---

[1] István Hargittai, *Our Lives: Encounters of a Scientist* (Budapest: Akadémiai Kiadó, 2004); I. Hargittai, *Wege zur Wissenschaft: Ein ungarischer Forscher berichtet* (translated into German by Manfred Stern, Freiburg: Lj-Verlag, 2006).

© Springer Nature Switzerland AG 2020
I. Hargittai, *Mosaic of a Scientific Life*, https://doi.org/10.1007/978-3-030-34766-6_14

Much of what I had to write about in connection with our family history was fading as time went by and I felt it important to have it recorded. This concerned especially the persecution of Jews. The current Hungarian officialdom does everything to whitewash the consistently anti-Semitic Horthy regime, but it is not only this right-wing political system that is complicit in falsifying history. The leftist-liberal media provide examples as well. Recently, there was an article in the literary magazine *Élet és Irodalom* (Life and Literature), which described the life in the Austrian labor camps (*lagers*) as almost idyllic in contrast with the experience of my family and others that is narrated meticulously in *Our Lives*. A reporter in the only remaining leftist-liberal radio channel characterized the present-day public work as similar to the infamous and murderous institution of the slave labor that conscripted Jewish men and boys had to perform during the first half of the 1940s. A socialist politician while in office as prime minister said that "we let them go" and "we were unable to protect them" referring to the half a million Jews who perished in extermination camps. Such statements by a prime minister of Hungary constitute falsification of history because the Jews were deported with the active participation of the Hungarian State. Speaking about the Holocaust, the language is often awkward and reflects a tendency of diminishing the enormity of the tragedy. Avram Hershko (see the Hershko chapter), the Hungarian-born Israeli scientist received the Nobel Prize in Chemistry in 2004. At the end of January 2005, he gave a talk at the Hungarian Academy of Sciences. The President of the Academy introduced him and he said, among others: ". . . He was born in Karcag . . . and he had to leave Karcag, because two thirds of the Jewish population was sent to other country and destroyed. He survived. He was one of many other Jewish colleagues who survived. . . ."[2] The euphemistic expressions of "leave Karcag" and "sent to other country" are unacceptable in the context they were used. *Our Lives* demonstrated what "leaving" our homes meant and, obviously, "other country" meant Auschwitz. The reference to "many other Jewish colleagues who survived" leaves me speechless.

Our Lives was translated into German and Russian. The German translation has appeared; there was no publisher to bring out the Russian translation. Curiously and by all indication, many more people have read the Russian translation than the German. Manfred Stern (1946–2018) of Halle, a mathematician turned translator, prepared the German version, mostly on the basis of the Hungarian text. Many years ago he was a postdoctoral fellow at the University of Debrecen and he learned perfect Hungarian. After he took

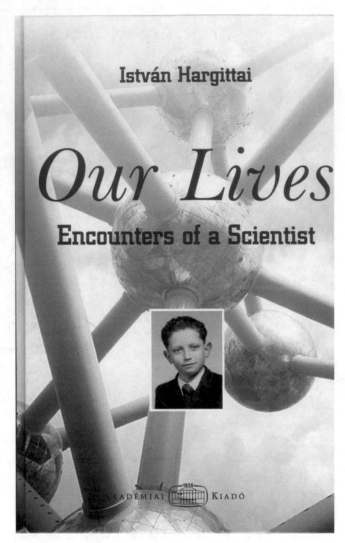

The cover of *Our Lives*.

early retirement as a mathematician, he founded a translation bureau in Halle. He worked on *Our Lives* with great care and spent some time with us while working on the translation. It is not only a good command of the languages that makes a good translator of a book like *Our Lives*. There is need to know history, to have a broad world view, empathy and humility as translator, and magnanimity in using the language. Stern possessed all these traits.

The circumstance for the Russian translation could not have been more different. I had known nothing about the Russian translation until I received it in a complete version. By then, it had been in circulation, mostly among Russian immigrants in Germany. Lately, it has become quite popular

---

[2] The quoted excerpt is from the English original: https://mta. videotorium.hu/hu/channels/9/vendegunk-avram-hershko-nobel-dijas-tudos (downloaded May 21, 2019).

Árpád Göncz and the author in 2006 at the annual festival of books (by unknown photographer).

in Russia proper as well. Thus, the book has had a substantial readership without having appeared in print.[3] I never met the translator, Erlen Fedin (1926–2009), but developed a correspondence with him. Professor Fedin was a physical chemist, and I knew his mentor, Aleksandr Kitaigorodsky, as well as some of Fedin's colleagues.

Fedin's parents were early Bolsheviks, dedicated to the Soviet order and named their son Erlen, from Era Lenina, i.e., Lenin's era. Erlen's father was executed in 1937, during Stalin's first Great Terror, and his mother was arrested, though eventually, they let her go. She was arrested again in 1949, during the next terror. At that time, Fedin was a physics student, and he was deported to a labor camp. Fedin could keep up with his studies due to a friend who obtained permission to send him his lecture notes. When in 1956 he and his parents were exonerated (his father only posthumously), he returned to Moscow. He had no certificate about his studies, but was determined to engage in science. Kitaigorodsky employed him and Fedin later referred to him as the first free human being he had ever met. Fedin was already 30 years old; he completed his studies and acquired his scientific degrees. By the time he retired he had been a

In dieser Schule waren 1944-45 etwa 500 ungarische Juden, darunter auch Kinder, interniert. Viele von Ihnen starben an den erlittenen Entbehrungen und Misshandlungen durch die Nationalsozialisten.

Gewidmet von der BVin des 12.Bezirkes
Gabriele Votava, 2006

---

[3] Added in fall 2019: Finally, the Russian translation has appeared in print: Иштван Харгиттаи, *Наши жизни: встречи учёного* (пер. с англ. Э.И. Федина; под ред. В.М. Тютюнника; Тамбов: МИНЦ, 2019).

Memorial plaque on the façade of the school at 10 Bischoffgasse in the 12th district of Vienna (photograph by the author).

Agnes Heller is presenting the Hungarian version of *New York Scientific* at a book launch, on June 7, 2017, at the Library of the Hungarian Academy of Sciences. The two authors, István and Magdolna Hargittai are also at the table (photograph by and courtesy of Klára Láng).

professor with an internationally renowned name in chemistry. He immigrated to Germany in 1995.

Our Lives has had quite some afterlife. During my writing it, I visited the school building in Vienna, which served as our *lager*, at 10 Bischoffgasse. The school officials met me courteously, but they knew nothing about the war-time history of the school. The school diary had only insignificant notes about the academic year 1944/45, and not a word about the *lager*. The English version of *Our Lives* reached the school and there was a suggestion to erect a memorial plaque on the façade. Nothing came of it for a while, but there was one person in the district leadership who did not let the idea die. The German text of the plaque in my translation is "In this school in 1944–45, some 500 Hungarian Jews, among them children, were incarcerated. Many died as a consequence of deprivation and abuse by the national socialists." An official of District 12, Gabriele Votava, had the plaque erected by private initiative, in 2006.

Agnes Heller read *Our Lives* in January 2018, many years after it was published. Few among contemporary Hungarians had such a grasp of the historical era covered in the book as she. From her letter to me: "I have read your book. Excellent.

It is well written and I understood even what I did not. It is a success story and your scientific persona, your persona, shines through best when you write about others. This is how you avoid the pitfall of most scientific success stories—boredom. Where Esterházy speaks about his 'fathers' you speak about 'brothers,' not blood brothers, but brothers in 20th century science."[4]

The philosopher Agnes Heller (1929–2019) was a disciple of George Lukacs. Her father was murdered in the Holocaust. Following the tragedy of Czechoslovakia in 1968, she could no longer believe in socialism as it was practiced in the Soviet camp. The Kádár regime forced her out of Hungary along with some other philosophers. She taught and researched in Australia and eventually became a professor at the New School in New York. She returned to Hungary after the political changes. She was a relentless critic of the autocratic regime building up in the 2010s. She was a prolific author and the intellectual atmosphere sizzled wherever she appeared.

Our Lives was the starting point for a number of new interactions of which I mention one more. I had met Oliver Sacks (1933–2015) even before the book had appeared; our

---

[4] E-mail message, January 9, 2018. Péter Esterházy (1950–2016) was a Hungarian writer.

István Hargittai and Oliver Sacks in 2003 in New York (photograph by Magdolna Hargittai).

correspondence grew following the publication. The American physician, neurologist, author, and science historian Sacks originated from a Lithuanian-Jewish family. He wrote bestsellers based on his experience in medical practice. One of his books, *Awakenings*, was made into a well-received movie in 1990, starring Robert De Niro and Robin Williams. Sacks found several of his friends among the heroes of *Our Lives*, such as Roald Hoffmann[5] and George Klein (see a separate chapter). Sacks singled out as of special interest to him what I wrote about growing up without a father. In this connection, I learned a great deal from him about the relationship between biography and autobiography, and I sensed his influence as I compiled the present volume.

---

[5] Roald Hoffmann (1937–) is a Nobel laureate (1981) chemistry professor emeritus at Cornell University. He writes poetry. Our friendship reaches back to the early 1970s.

A Memorable Nameless in Science

József Hernádi and the author in fall 1966 in Visegrád during an outing of the Research Laboratory of Structural Chemistry (by unknown photographer).

I have had this recurring sensation that I met people when I needed them whether it was a teacher, a professor, a mechanical engineer, and so on; the right person at the right time. Among such lucky meetings, my association with József Hernádi was one of the luckiest. He was in charge of the mechanical workshop of our Laboratory. He had four junior associates working in his team.

Today's research in much of chemistry is done with apparatus commercially available; you "only" need funds to acquire them. The truly well-to-do laboratories though build their own experiments and that is where real innovation comes from and that is where there is a higher probability of making discoveries. In the second half of the 1960s, we did not have funds to purchase commercial instrumentation, but

© Springer Nature Switzerland AG 2020

I. Hargittai, *Mosaic of a Scientific Life*, https://doi.org/10.1007/978-3-030-34766-6_15

we had this first-class mechanical workshop. Such experts today would be impossible to hire in academia. In today's Hungary, they could not be paid the high salaries they have every right to demand and many of them may have already left academia seeking employment in other countries. In the 1960s, there was no mobility and an academic laboratory offered many advantages, for example, flexible hours and challenging projects. Unfortunately (though fortunately for me, as it turned out), most associates of our Laboratory asked the workshop to prepare meaningless apparatus—it was a sign of prestige to keep the workshop busy with your tasks. Hernádi went along because he could not declare his workshop superfluous.

My joining the Laboratory changed the situation; his workshop started producing innovative pieces, which we published together with Hernádi—he had never had any publication before. We built a semiautomatic system for introducing the vapor beam to cross the electron beam within the diffraction space. An American professor on a visit saw it and asked us to prepare a copy for him. He said we would come back next year to pick it up. We did and he did. We could not sell it to him; there was no mechanism for such trade, but he gave us a handheld calculator, which was the first in our Laboratory.

Hernádi was in his early fifties during our joint work, which lasted a mere 3 years. Our joint output though was many times over what usually happens in 3 years' time. Initially, I had the opportunity of modifying an electron microscope for my experiments. Several research groups used this instrument; the hours of operation had to be shared and I became an additional user. I had the instrument for 1 day weekly. The other groups used it for routine purposes.

They brought their samples, recorded their images, and left. We did something entirely different. I had to rebuild the apparatus before I could start my experiments. Hernádi "lent" one of his associates to me and we started work early in the morning, removing parts of the apparatus, and building in the parts that were being developed in the workshop. Every week we carried the modification of the original instrument one step further. All had to be done in such a way that the next morning the apparatus was ready for the next research group's routine experiments. Usually, we finished at dawn, just a few hours before the others came; and they did not even notice the temporary metamorphosis of the apparatus. It was a modern Cinderella story.

By the time Hernádi had passed away (a terrible illness finished him off in a few weeks' time) our research group had already developed an international reputation for innovation. Besides, I learned from him not to be afraid to consider entirely different solutions from what others might have done. Somehow, we kept finding qualified mechanical engineers for many years. I like to think that the creative atmosphere may have attracted them—earning possibilities could have not, that is for sure. Due to this reputation, we had many visitors, including long-term visitors, which was highly unusual in the divided world at the time. From Eastern Europe, many were eager to go to the West for months and years let alone those that never returned. From the West, it was a rarity that scientists would come longer than a few days or weeks. We had Italians for multiple three-month stays; Norwegians spending their sabbaticals; Americans for 1-year stays; British to complete their PhD studies; and so on. Curiously, we had our first doctoral students from Russia only after the political changes of 1989/90.

We Were in the Same Transport

The two photographs differed in a few months and a Nobel Prize. Top: Avram Hershko and the author in August 2004 in Woods Hole, Massachusetts (photograph by Magdolna Hargittai). Bottom: January 2005 at the Hungarian Academy of Sciences; from left to right: Magdolna Hargittai, Avram Hershko, Judy Hershko, and István Hargittai (by unknown photographer). Hershko's Nobel Prize was announced in October 2004.

© Springer Nature Switzerland AG 2020

53

I. Hargittai, *Mosaic of a Scientific Life*, https://doi.org/10.1007/978-3-030-34766-6_16

Avram Hershko (1937–) received a share of the 2004 chemistry Nobel Prize for the discovery of the ubiquitin-mediated degradation of proteins. Ubiquitin is a protein which is present everywhere in the organism. For maintaining life, it is necessary not only to create but also to destroy proteins. We visited the Hershkos in August 2004 in Woods Hole, a well-known resort place in Eastern Massachusetts on the Atlantic Ocean. It is also a renowned research center, the venue of the Oceanographic Institution. Hershko had been renting lab space at this location for years. This was our first meeting and we recorded a long conversation for our *Candid Science* series.

Hershko was born as Ferenc Herskó in Karcag, a small town in eastern Hungary of about 25 thousand inhabitants with about a thousand Jews before the Holocaust. Today, there are hardly any Jews left. In spring 1944 the Jews were first gathered in a ghetto in Karcag and Hershko remembers the brutality of the Hungarian gendarmes ordering them out of their homes. They were then moved to the ghetto in Szolnok, a larger town where the conditions were especially harsh. It was an open-air camp exposed to rain and thunderstorms and nothing to sleep on in the night. Some tried to escape and the gendarmes beat them ruthlessly in front of the children. Hershko was 6, his brother 8.

Next, they were put into cattle carriages and the train started moving. Some family members were on a different train and ended up in Auschwitz. Hershko's train ended up in the Austrian distribution camp in Strasshof an der Nordbahn, a suburban town of Vienna. They spent the next months in a smaller labor camp in the village Guntramsdorf, near Vienna. For all we know, we may have been in the same train (more about it in the Fejes Tóth chapter). Even though the conditions were harsh here too, it was not as brutal as the ghetto in Szolnok. Upon liberation, in April 1945, they started back on foot to Hungary. They found their house in Karcag empty, and nothing was returned to them voluntarily. Hershko's mother went around the neighboring homes and recognized some of the furniture that was then brought back to them.

Hershko's father had been taken to a slave labor camp and to the Russian front. Most of the Hungarian guards were tolerable, but a Hungarian army major ordered them to strip and run in the snow. To the father's luck, the Russians captured him. He spent years as a POW and returned to Hungary in 1947 and wrote up his experience. He was bitter that not a single Jewish family was saved in any of the farmhouses in the vast farmland around Karcag where they could have found appropriate hiding places. Hershko's maternal grandparents were murdered in Auschwitz and so was his four-year-old niece. Her mother, Hershko's aunt, used to have beautiful black hair; when she came back her hair was white.

Hershko's father used to be a teacher in the Jewish school, but after the war, there were no children for him to teach in Karcag. They first moved to Budapest, then, in 1950, to Israel. The father became a successful pedagogue. He taught in a teachers' college and wrote textbooks. He lived to 93 years and Hershko's mother to 91. Hershko's brother is a renowned professor of medicine in Israel.

Hershko first graduated from medical school, then earned a PhD degree in biochemistry and dedicated himself to research. He chose the project of protein degradation while doing postdoctoral studies in the USA. At the time most scientists in the field were interested in the formation of proteins, but Hershko never wanted to go with the crowd. He had two superb mentors from whom he learned a great deal, but it was equally meaningful that he observed behaviors that he later wanted to avoid. One had the tendency of doing too many things at the same time and thus wasting his talent. The other, although he was a physician by training, neglected his illness, and died. There was then a third mentor with whom he was on equal standing, Irwin Rose (1926–2015), and who then became one of his two co-laureates. The other co-laureate was one of Hershko's former disciples, Aaron Ciechanover (1947–). The real significance in the discoveries of Hershko and his colleagues was in understanding the degradation of proteins because that might lead to finding the means of degrading the proteins of cancerous cells.

Aaron Ciechanover (top) and György (George) Konrád (bottom) and the author in May 2005 in Budapest (photographs by Magdolna Hargittai).

The beginning of my interactions with Aaron Ciechanover also dates back to a time before the Nobel Prize. We first met in 2003 at a meeting of the graduate students of the Karolinska Institute in Sweden, on a small island of the Stockholm archipelago where I gave a talk about success in science. Ciechanover is an easy person to talk with, very different from his former mentor. Hershko does not enjoy the limelight; Ciechanover does not mind being a celebrity.

When we visited both in Israel, Ciechanover showed us the special room in his home where he keeps his trophies. Hershko took us to his circle of friends where he is emphatically only one of them.

Ciechanover is deeply interested in Jewish tradition and he enjoyed immersing himself in the remnants of Jewish life in Budapest during his visit in 2005. He discussed with the renowned author György (George) Konrád (1933–2019) the dangers of disappearance of those relics. It was also in seeking out surviving traditions that he met with the poet András Mezei (1930–2008) and visited the Scheiber Sándor High School and General School.

Bestseller from Hobby

Our most successful book on symmetry with the image and its reflection of the Sant' Angelo bridge over the river Tevere in Rome (photograph by the author).

© Springer Nature Switzerland AG 2020

57

I. Hargittai, *Mosaic of a Scientific Life*, https://doi.org/10.1007/978-3-030-34766-6_17

Our Hungarian book about symmetry for children with lots of images appeared in 1989. We were close to the political changes, but state subsidy for book publishing still existed. The book sold out quickly and additional printing was not possible due to the disappearance of state support. The book was exhibited at the Frankfurt Book Fair where it caught the eye of an American publisher, Lloyd Kahn (1935–). His company, Shelter Publications in Bolinas, California, was a one-person enterprise. Lloyd was the owner, editor in chief, and most employees were family members. Our interactions brought us a unique publishing experience.

Lloyd sold enormous numbers of copies of his books that were about health, weight lifting, exercising, the design and maintenance of septic tanks, how to build your own dwelling, and suchlike. Self-designing houses was his favorite topic and the name of his company reflected this. By looking at his list it would have never occurred to us that he might publish our book. He did not expect commercial success from the experts; and if there was a choice in solving any technical problem, he always opted for the best.

The book appeared in 1994 and became an instant success. *Scientific American* and other well-known publications praised it. Lloyd collected quotes from opinion makers and they were reproduced in further printings. For a while, Random House was the distributor rather than Lloyd's usual partners. An abridged German edition also appeared.

Lloyd lives in his self-designed house in Bolinas and they make a perfect couple. Artists and other private people live in Bolinas who do not seek publicity; on the contrary, they do what they can to avoid it. From time to time they remove the sign of the town so it may be difficult for travelers to find it. However, the people are most friendly to those who come there on legitimate purpose and not just for taking snapshots of the "natives." On our visits, we stayed in a bed and breakfast; people knew we worked with Lloyd and made us feel most welcome.

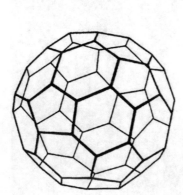

The truncated icosahedral model of the C_{60} molecule—buckminsterfullerene (The truncated icosahedron is obtained by symmetrically shaving off the corners of the regular icosahedron. The latter has 20 identical regular triangular faces.) and the Geodesic Dome in Montreal (detail) in 1995. One of the pentagons among the hexagons is marked (photograph by Magdolna Hargittai).

symmetry book, but he liked it and thought that for once, he could publish a book for hobby. He did not think it would sell as a children's book in America, so this aspect was dropped.

It took years to produce our book; both Magdi and I spent a week or so in Bolinas on separate occasions. Lloyd worked on the book with gusto; designed each page; consulted with

Lloyd was serious about architecture and understood that it is not only the external view that counts but the structure must also be stable statically. In his youth, he was a follower of R. Buckminster Fuller (1895–1983), who designed the famous Montreal Dome. When a group of chemists and physicists discovered the beautiful C_{60} molecule whose geometry resembled the Montreal Dome, they named the molecule buckminsterfullerene. The discovery of a whole class of molecules, called fullerenes, followed. They are characterized by a spherical shape in which there may be any number (except one) of regular hexagons and 12 regular pentagons. When we had planned our Montreal visit with the specific purpose of looking at Fuller's Dome, we had almost been talked out of it for it had burned down. In reality, only the outer plastic shell had burned, making the underlying steel structure—the one we were interested in—yet more visible.

I never met Buckminster Fuller, but had friendly interactions with his former close associate, Edgar (Ed) J. Applewhite (1919–2005). Our common interest in geometry brought us together. He helped Fuller in producing Fuller's opus magnum, *Synergetics*, but stayed in the background. Ed had served in the CIA for a quarter century before joining Fuller, who was known for accepting the contributions of others without giving them credit. Ed told me about their interactions and added that had he (Ed) craved publicity, he would not have chosen to work in intelligence. Ed had two copies of the two-volume *Synergetics*. First, he gave me one copy and a few years later the second copy. He was already terminally ill, and the next time he told me that it was our last meeting; he wanted to part when we could still be having a meaningful exchange.

Edgar J. Applewhite in 1997 in his home in Washington, DC (photograph by the author).

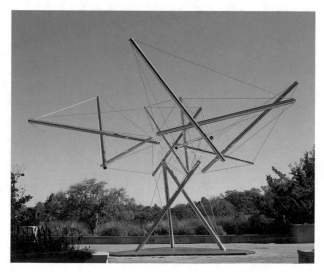

Kenneth Snelson in 1997 in his New York workshop and one of his tensegrity sculptures in front of the National Library of Medicine in Bethesda, Maryland (photographs by the author).

Not all of Fuller's disciples were happy remaining nameless. Fuller liked to discuss constructions that were held together by tension between their components, called *tensegrity*. The sculptor, Kenneth Snelson (1927–2016), discovered these structures and he once lamented that Fuller failed to mention him in his discussions. Fuller responded that Snelson was young enough for staying without recognition for a while. Ken was an artist; he was different from Ed Applewhite—Ken did crave public recognition. His sculptures stand at prominent locations.

Innovative Design

Gyorgy Kepes in 1984 in his home in Cambridge, Massachusetts (photograph by the author).

My interactions with Gyorgy Kepes (1906–2001) followed from my first edited volume on symmetry. He immigrated to America from Hungary in 1937 and spelled his first name as Gyorgy. He was a design scientist and artist, wrote much about art theory, and was Professor of Design at the Massachusetts Institute of Technology (MIT). His book series, *Vision + Value*, served as a model when I embarked on organizing the *Symmetry: Unifying Human Understanding* volumes.[1] At the time, I was a visiting professor at the University of Connecticut in 1983–1985, the first year in the Department of Physics and the second year in the Department of Chemistry.

I invited Kepes for a contribution to the symmetry project. He responded kindly, but he found it too hard to prepare a chapter for the book. Instead, he invited me for a visit to Cambridge, Massachusetts. For American distances, we were almost neighbors and I was happy to go.

---

[1] István Hargittai, Ed., *Symmetry: Unifying Human Understanding* (Oxford: Pergamon Press 1986). István Hargittai, Ed., *Symmetry 2: Unifying Human Understanding* (Oxford: Pergamon Press 1989).

© Springer Nature Switzerland AG 2020
I. Hargittai, *Mosaic of a Scientific Life*, https://doi.org/10.1007/978-3-030-34766-6_18

Two Kepes photographs. Left: "Dendric liquid" (1949); right: "Dendric network" (1962). Courtesy of the late Gyorgy Kepes.

He lived in a big house surrounded by a large garden, and was experimenting with the largest ever Polaroid camera lent to him by the company. He gave me a few of his extraordinary pictures for free utilization in my symmetry volumes. They fit eminently my loose interpretation of what symmetry was.

The idea for producing the *Symmetry* volumes came from Ervin Y. Rodin, the Hungarian-American editor of the periodical *Computers and Mathematics with Applications*. Rodin lived in St. Louis, Missouri, and knew me only from my publications. The collections first appeared in this journal and subsequently as books. The two volumes comprised over 2000 large-format pages with contributions by artists and scientists and have become collector's items. For me the most valuable benefit from this project was the meeting of exceptional minds, such as Kepes (some others have already been mentioned).

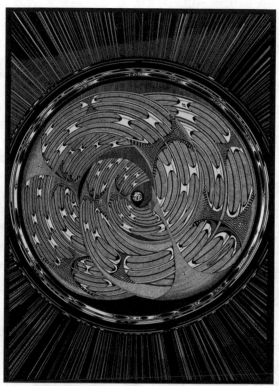

Left: Anatoly Fomenko and the author in 1987 in Fomenko's Moscow home (courtesy of Tatiana Fomenko).
Right: "Visible and hidden symmetries in geometry" by Anatoly Fomenko (courtesy of Anatoly Fomenko).
The Publisher used this image on a big poster popularizing the second volume.

The world-renowned Russian mathematician, Anatoly T. Fomenko (1945–), was one of the many contributors to these *Symmetry* volumes. Sometime in the mid-1980s, during a visit to Moscow, I noticed some unusual photographs on sale by street vendors. They were of large size, black-and-white, powerful, and abstract. My eyes, sensitive for noticing symmetries, recognized plenty of them in these pictures. Their abstract nature, however, was intriguing. The sellers had no idea about their author and they were not at all eager to engage in any conversation about the origin of their wares. Abstract art was illegal in the Soviet Union.

Eventually, I found out that a mathematics professor of Lomonosov University, Anatoly Fomenko, was the artist of the graphics and that he was not aware of the availability of his art on the street. Incidentally, not only then, but when computerization had become so much more accessible, he continued producing his art by hand.

Fomenko's contribution to the first edited symmetry volume was on the mathematical description of soap bubbles.

His peculiar graphics appeared in the second volume. They represented mathematical concepts. I asked him if he had ever considered that the authorities might take action against him in view of the illegality of abstract art. He merely illustrated mathematical concepts, he said.

His graphics might be considered somewhat alienating though the frustration is eased by the presence of some humans in his pictures. Alas, these humans are almost invariable tiny as compared with the principal motives so they add to the perception of hopelessness if one is inclined to having such feelings.

Fomenko's international fame originates from three sources: his publications in mathematics; his graphics; and his books on the general history of mankind. He popularized a new historical time scale. He proposes it to replace the supposedly false time scale that had ostensibly been the creation of historians and that omits a few centuries from human history. In our personal meetings, the latest was in September 2016, this topic never came up.

Werner Witschi and one of his moiré sculptures (photograph by the author).

The Swiss sculptor Werner Witschi (1906–1999) was another among the contributors to the *Symmetry* volumes. His home and workshop were in Bollingen, near Bern, and Magdi and I visited him a few times. He started out as a traditional artist, but his fame came eventually from his kinetic art representing moiré patterns. This is what brought us together. His sculptures represented a great deal of symmetry and their symmetries had the ability to change due to the interferences inherent in his moirés. Such patterns can be created by the superposition of the most rudimentary planar patterns. Witschi had a mathematician friend, Hans Giger, who produced a mathematical description of Witschi's moirés and we communicated his paper along with Witschi's writing, which was copiously illustrated with his art.

Still-life by Károly Kerti from the 1950s (photograph by the author).

Károly Kerti (1917–1986) was my teacher of drawing and geometry—two separate subjects—in general school in Orosháza. Kerti was more an artist than a pedagogue. Teaching was a burden for him and children recognize this quickly. Kerti had an exhibition in our town and Mother bought one of his pieces, a still life with sunflowers. This was when I was not his student anymore.

Kerti was a latecomer in art because as soon as he had graduated from art school he was conscripted, and during most of World War II, he was a prisoner of war. He returned to Hungary only in 1947. Originally he was to become a painter, but due to poor finances, he moved to drawing. He exhibited in many countries and received prizes. He could not teach me to draw, but I liked geometry.

* * *

© Springer Nature Switzerland AG 2020

I. Hargittai, *Mosaic of a Scientific Life*, https://doi.org/10.1007/978-3-030-34766-6_19

Fairy tale world of János Balázs from the 1970s (photograph by the author).

János Balázs (1905–1986) was a gypsy painter about whom I read in a women's magazine in the early 1970s. His works were so-called "primitive art." He lived in a gypsy community, one could call it a ghetto, in the outskirts of Salgótarján, a mining and industrial town, 100 km (70 miles) to the north-east of Budapest. I visited him with a friend of mine. That he lived in poverty is an understatement; he had nothing except his paintings and he painted on every material he could lay his hands on. There was meticulous cleanness in his shack. I would have liked to buy a picture from him, but I did not even try. I could not have paid him an adequate price and later I heard that he did not sell his pictures. There was one exception, when he wanted to have his poetry published, and sold some of his pictures to pay the press. Once he had been "discovered," the authorities moved him to an apartment in downtown and he was subjected to learning painting. As a result, he no longer produced anything. It was years after he died that I noticed one of his paintings in a shop of old books and I bought it.

* * *

Fall melancholy by János Halápy (photograph by the author).

We were a happy family of four at the time of my birth in 1941. We lived in a two-story building of our own on a large lot at 9 Bécsi (Vienna) Avenue in the Third District of Budapest. The building still stands on a much smaller lot. My parents rented out the basement and the ground floor and we lived upstairs. Part of our home was Father's law offices. Soon he was called for slave labor as were most Jewish men. He was sent to the Eastern front and we received the news about his death in the fall of 1942. It was then that Mother, my brother, and I moved to Orosháza. When we returned to Hungary from the deportation, we first went to Budapest. We found our home occupied by a Soviet command post. We could take only some of our books with us; everything else was being used or had disappeared. The Halápy painting shown here used to be in our home and I have no idea how we saved it. I always liked this picture for its melancholy and for the fact that it was from our original home from which so little has remained in our possession. János Halápy (1893–1960) was the painter of Lake Balaton, but this picture is not one of his characteristic paintings. We received it from the artist in lieu of some lawyer services rendered by my father to the artist. Many years ago, I found a book about Halápy's art with a reproduction of this painting with a note that it was "hiding." We sought out Mrs. Halápy and she was happy to visit the picture and know that it was no longer "hiding."

* * *

I was a Visiting Professor at the University of Connecticut (UConn), 1983–1985. UConn often organized 1-day bus tours to New York City and we participated whenever we could. The bus discharged us on the elegant Upper East Side of Manhattan and picked us up at the same place. The location was between the Metropolitan Museum and the Guggenheim Museum. We walked around a great deal and visited small galleries too, not only the great museums. The name of the owner of one of the galleries sounded Hungarian and we made his acquaintance. He was Paul Kovesdy—Pál Kövesdy (1923–2017). He was more an art collector than a businessman and he invited us to his home nearby. It was a small apartment with all its walls covered by Hungarian

art collector. In his youth, he used to be aimless. He was taken to a slave labor camp because he was Jewish; he suffered physically and spiritually at the hands of ruthless guards, but survived. At one time he wanted to become a singer, but it did not happen. After the suppression of the 1956 Revolution, he became a refugee and moved to the USA. He became the proprietor of a small gallery by accident, but then he learned to love the Hungarian avant-garde. It surprised us how much we learned about him, and he told us that our children's visible interest opened him up. Following the political changes of 1989/90, he moved back to Budapest, lived off his collection, and died lonely.

* * *

In hindsight, it is sad that there was no communication between Károly Kerti, the artist and art teacher, and Károly Kerti, the geometry teacher. Their dialog might have benefited his teaching of both subjects. My interest in geometry may explain my love of avant-garde art, but the geometry I am interested in is not the one of mathematical rigor. Understanding the transition from rigorous geometry to the avant-garde art might help gain insight into both. I am trying to illustrate the meaning of the removal of geometrical rigor with an example.

"The Sphere," Franz Koenig's sculpture in the southern tip of Manhattan, in the 1980s, on the left; and in 2014, on the right (photographs by the author).

avant-garde paintings. We thought he must have been immensely rich, but he was not. He had bought all his pictures dirt cheap in Hungary at the time when this art was not yet appreciated. The Hungarian avant-garde made him an

There are an infinite number of symmetry elements characterizing the sphere, but only the perfect sphere, which is the geometrical sphere. If we damage the sphere in any way, it loses its symmetries; it is no longer a sphere, but only in the sense of geometry. Otherwise, it remains a sphere—to us. A sphere would have to be damaged quite severely to be no longer recognized as a sphere. When we took a snapshot of Franz Koenig's (1924–2017) sculpture, "The Sphere," sometime in the early 1980s at the World Trade Center in Manhattan, it did not bother us that it was not perfect. The terror attack on September 11, 2001, destroyed Koenig's sculpture along with thousands of human lives. The sculpture could be reconstructed to some extent from the parts found in the rubble. This reconstructed version is even farther from the perfect sphere though no one doubts that it still represents a sphere. It still carries the name "The Sphere" and has metamorphosed from a sculpture into a memorial.

George Klein

Great Klein

Georg Klein in 2000 in the Hargittais' home in Budapest (photograph by the author).

The Hungarian György Klein was born on July 28, 1925, in Budapest and the Swedish Georg Klein died on December 10, 2016, in Stockholm. He was one of the best-known cancer researchers and immunologists of his time. According to some, the only reason he did not become a Nobel laureate was that for many years he was a leading member of the Nobel Assembly of the Karolinska Institute responsible for the Nobel Prize in Physiology or Medicine.

We first met in 1996 and then had many conversations during the ensuing years, some of which we recorded. We met in Stockholm, Budapest, and the USA. One of his favorite topics was death, which he considered a natural component of life. It would be better to prepare for it, but in most cases only an incurable illness makes us think about the finiteness of life. Another of his favorite topics was conformism. I only listened to what he had to say about death, whereas conformism was something I was also deeply interested in—more about it later.

The Jewish Klein started his medical studies only after the liberation of Hungary in 1945, a 2 year delay, first in Szeged then he continued in Budapest. He sensed early on the approaching communist dictatorship and in 1947 he left Hungary for Sweden. He completed his training in medicine while doing research at the Karolinska Institute between 1947 and 1949 and received his MD diploma in 1950. He was an associate in the Karolinska from 1951 and was the founding director of the Institute of Tumor Biology and Cell Biology of the Karolinska until 1993. Although the Swedish regulations for retirement are rigorous, as an exception, he could continue as a group leader for the rest of his life.

Klein and his wife, Eva Klein (née Fischer, 1925–), worked together in science throughout their careers, but developed their independent lines of research as well. When travel programs separated them, they were on the phone all the time calling one another—I witnessed this on more than one occasion. They discussed the minutest details of their activities. Eva was also given the privilege to keep working at the Karolinska following her retirement.

The young Klein burst into the Swedish scientific life. His main activities focused on experimental cancer research and tumor immunology. Although he was an experimentalist, he considered his most salient contribution to science an insight

---

A Budapest girlfriend Called Klein [klein = small] "groβ" [big].

I. Hargittai, *Mosaic of a Scientific Life*, https://doi.org/10.1007/978-3-030-34766-6_20

rather than an experimental finding. This insight was that chromosomal translocations in tumors, that is, exchanges between two different chromosomes, must reflect oncogene activation effects. This may sound esoteric, but it helped a great deal to understand the origin of cancerous developments in the organism. Klein could also formulate the starting points of his research in the simplest, most accessible way. Thus, for example, he wondered that if one of three people has cancer, that person becomes the target of investigation. However, it is also true that two of three people do not have cancer. Understanding why they do not have cancer may bring us closer to avoiding cancer and treating it when it happens.

Klein participated in making decisions about the Nobel Prize in Physiology or Medicine for 36 years. Thus, his opinion is worthwhile to consider even though it was extreme, might even say, it was deliberately shocking. He was irritated by what he considered the out of proportion prestige of the Nobel Prize. He called it the illness of *Nobelitis* when somebody who was close to the possibility of getting the prize became obsessed with this possibility. A true researcher would not succumb to such an illness—he opined. The other side of this phenomenon was *Nobelomania*, the obsession of those that considered the importance of Nobel Prize above everything else and they included some of the members of the prize-awarding bodies. Klein intended to bring down the Nobel Prize from its pedestal and stressed that the Nobel laureates may be as stupid as anybody else. This is not only shocking, but is also valid with a marked exception. A Nobel laureate may be as ignorant as anybody else in everything with the exception of the area where he/she made the prize-deserving contribution. There, the laureate may be the smartest person in the world or one of the smartest. I would emphasize the sad watershed effect of the Nobel distinction. The names of those who receive it are accorded immortality of sorts, while the names of those who do not, however close they had been to getting it, may disappear into oblivion, even if their contributions were commensurable with those of the laureates.

Klein was almost 60 years old when he embarked on authoring books. He wrote in Swedish and most of his books have also appeared in English and Hungarian translation. The Holocaust and anti-Semitism in Hungary figured often in his writing. Klein's uncle was Prime Minister Pál Teleki's physician. Klein was still in high school when this uncle asked Teleki for help in getting Klein enrolled in medical school. By that time the *numerus clausus* legislation restricting Jewish youth from getting a higher education was becoming *numerus nullus*. Not only did Teleki decline to help; he warned his doctor not to mention his nephew's name. If he had learned the name, Teleki said, he would personally prevent him from entering medical school. Teleki was fiercely anti-Semitic. By the time Klein might have applied to medical school, the question was no longer whether or not he could study; the question was whether or not he would survive.

It was still under Nicholas Horthy's reign that the entire Jewish population of Hungary, with the exception of Budapest, was deported to Auschwitz. It was done as if the deportees were merely relocated and many believed this—the reality was too unbelievable. Klein at the time was working for the Budapest Jewish Council and he had access to the document having become known as the *Auschwitz Protocol*. It was compiled by two Slovakian Jews, Rudolf Vrba and Alfred Wetzler, who escaped from Auschwitz. They warned the Hungarian Jews of what they were facing in Auschwitz. Having read this document, Klein escaped from a transport destined for Auschwitz. Klein had been asked by reporters and politicians how he imagined his career might have been had he stayed in Hungary. Expatriate Hungarian scientists often have to respond to such questions. Politeness dictates that the respondent evades telling the truth. Those who pose the question know the truth yet pose the question anyway counting on the respondent's willingness to collude in evasion. It is the most sanctimonious game playing.

As I mentioned in the introduction, conformism was a favorite topic in our conversations. I started reading serious books early as a child. I read repeatedly Heinrich Mann's *Der Untertan* (*The Loyal Subject*) and W. Somerset Maugham's *The Razor's Edge*. These books told me a great deal about various occurrences of conformism. In *Der Untertan*, it was about the determination of Imperial Germany to replace individualism by having people give up their own personality. In *The Razor's Edge*, it was about an American youth desperately looking for an escape from the societal pressure forcing on everybody a generally accepted lifestyle and behavior.

Klein compared the societal atmosphere in Hungary, Sweden, and the USA. As soon as he could leave his hiding place following the liberation of Budapest in spring 1945, one of his relatives took him to a communist party meeting. On their way to the meeting, Klein witnessed how his relative was instructing his wife as to what she could and could not be saying at the meeting. It was clear to Klein that in the anticipated communist dictatorship, everything will be coming from the top and people will have to live with an arrangement in which others will make decisions impacting their lives. He decided that he could not live under such conditions. In Sweden, he was happy that he did not need to be an accommodating member of society and did not need adapting his behavior to gain the approval of his neighbors. His neighbors may have laughed at the peculiarities of his behavior, but they respected his individualism. In the USA, he experienced diminishing tolerance and enhanced pressure by society to conform to the expected behavior. He knew it would be a recipe for his unhappiness. All this was not an abstract thought experiment for him. In 1966 he received a prestigious invitation from Harvard University to become Professor of Genetics with most favorable conditions and superb students. He was deliberating for 6 months when it finally dawned upon him that he would not accept it, he never seriously considered it, and the whole exercise was a performance for the galleries. He could not have become a "good American" because he placed too much value on his personal independence. He realized that this independence was a decisive motivation for him from his childhood in which the early loss of his father played a major role. Placing such a value on personal independence was the foundation of our friendship.

* * *

Prize in Literature and borrowed my book, *The Road to Stockholm*. He received the Nobel in 2002, during their Berlin stay.

One of the Hungarian TV channels gave a live broadcast of the Nobel award ceremony from Stockholm. In the Budapest studio, my friend, the esthetician Sándor Radnóti was asked to comment on the literature prize, and my task was to reflect on the rest of the prizes. The Hungarian reporter in Stockholm was doing a good job except that he did not distinguish between two Hungarian expressions of the word

From left to right: Hungarian composer György Ligeti (1923–2006), Eva Klein, Georg Klein, Imre Kertész, and Albina Vas (Mrs. Kertész). By unknown photographer; courtesy of Eva Klein.

The Kleins introduced us to Imre Kertész (1929–2016) long before his Nobel Prize. After the death of Albina Vas, his first wife, Kertész married the Hungarian-American Magda Sass. At some time Magda was the trade representative of the State of Illinois in Hungary and had a busy societal life. When Imre was required to accompany her, he often asked me to come along to keep him company. Our last meeting was at the end of August 2002, on the eve of his one-year Berlin scholarship, and we had them in our home for dinner. By then they were seriously expecting his Nobel

"flag," one that is common and another that is more solemn. The verbatim translation of the title of the novella "The English flag," which is an inadequate translation for it fails to transmit the solemnity of the original. The English translation has appeared under the title *The Union Jack*. This is exact though I wonder if *The British Colors* might have not been closer to expressing the atmosphere Kertész was creating. *The Union Jack* is my favorite among the Kertész oeuvre. On one of the more peaceful days during the brief period of the October 1956 Revolution, a car is speeding along the Grand Boulevard in Budapest, probably toward the airport, with a Union Jack draped over its radiator. The people on the boulevard, walking along in the fall sunshine, applaud as the car is passing by. As if in response to the applause, a gloved hand appears in one of the car's windows on the left side of the car, and the hand is waving from the car. This waving may be taken as an expression of solidarity. It may also be taken as an expression of sorrow as the owner of the gloved hand will presumably soon land in a London airport, whereas the people being left behind are facing an uncertain future. This brief scene appears at the end of the fairly long novella. The narrative is moving unhurriedly toward the concluding scene. Yet the reader does not get tired of the seemingly out of proportion introduction, which is full of concise description of a period in Hungarian reality. When I first read it, I felt a strong déjà vu because I used to walk along the Grand Boulevard in those days. I was a high school student and lived in a rented room not far from it. There were no instructions in the schools, every day carried an uncertainty of what the next day will bring, and the weather was most friendly. My déjà vu was that I remembered this scene or if not exactly this, something very similar. Probably, I did not witness it, but in Kertész's description, I was there, as if one of those on the sidewalk. I am not sure if I had joined in the applause; I seem to remember that I did not because I was too reserved for that.

No Role Model But Tremendous Impact

Statue of Arthur Koestler by Imre Varga on Lövölde Square, District VI, not far from Koestler's childhood home (photograph by the author).

Arthur Koestler (1905–1983) wrote one of the most compelling political novels of the twentieth century, *Darkness at Noon*. It first appeared in 1940. He was the first who unmasked Stalin's terror using literary means, including the show trials. The book was a powerful weapon against totalitarianism. Another writer unmasking the idea of totalitarian regimes, George Orwell, benefited from Koestler's experience. Orwell understood in the Spanish Civil War to what a large extent the media falsified reality and turned it into propaganda. This is why he stated that history stopped in 1936. Of course, it did not.

I have met two prominent individuals who stressed the impact of *Darkness at Noon* on the formation of their political views. One was Edward Teller about whom there is a separate chapter. The other was Martin Gardner (1914–2010), the well-known disseminator of popular science who edited the column "Mathematical Games" in *Scientific American* for 25 years. Gardner embarked on studying

© Springer Nature Switzerland AG 2020
I. Hargittai, *Mosaic of a Scientific Life*, https://doi.org/10.1007/978-3-030-34766-6_21

philosophy, but he was always interested in puzzles and while working for *Scientific American* he built up close interactions with mathematicians. Many thought he was a mathematician.

to write a negative review about it. He published the review in a reputed journal under a pseudonym. Although he included his own name at the end of the review, the book sales plummeted. This was not the only occasion when

Martin Gardner with the author in fall 1996 in Asheville, North Carolina, demonstrating his magic (photograph by Magdolna Hargittai).

Koestler read Gardner's column regularly and they corresponded. When reading the column, Koestler felt as if playing chess with a master. Their views on parapsychology, however, diverged. Koestler believed in it, Gardner thought it was unscientific. When we visited Gardner in 1996, he showed us some tricks from his repertoire as a magician. What brought us together was our shared interest in symmetry. He wrote about one of our books (see the Lloyd Kahn chapter) in a beautiful way.[1] He was extraordinary as the next story suggests. He published a philosophy book in 1983, which was selling well, when Gardner challenged himself

Gardner provoked his readership. He published a poem by the modernist American poet, William Carlos Williams, along with its parody and asked his readers to evaluate the two. The readers judged the parody to be far inferior to the original. Only then Gardner revealed that what he labeled "parody" was indeed the authentic one.

Returning to Koestler, he was born in Budapest and emigrated like so many others at the time. It was not only because of anti-Semitism but also because of the general hopelessness of the state of affairs in the Hungary of the 1920s. He became a communist and a significant author and reporter writing in the German language. Later, he switched his language to English, became a fierce anti-communist and a bestselling writer. He was a most original, creative mind and one of the giant contributors to twentieth-

---

[1] Martin Gardner, *The Night Is Large: Collected Essays*, 1938–1995 (New York: St. Martin's Press, 1996), p. 3.

century world culture. However, I would not recommend him for a role model. The way he treated women was despicable.

Koestler did not figure in my readings as he was nonexistent for socialist Hungary. In the academic year 1983/84, I was visiting professor at the physics department of the University of Connecticut at Storrs. Being a physical chemist, it was a challenge for me to teach physics. In my office, I found books left behind by my predecessor, among them Koestler's *The Sleepwalkers*. In contrast to the rather dull physics text I was using for the curriculum, it was a refreshing read. I could inject some of Koestler's stories into my lectures.

Johannes Kepler is the central hero in *The Sleepwalkers* and there is a resemblance between Koestler and Kepler. The sections of *The Sleepwalkers* about Kepler were published as a separate volume under the title *The Watershed*. Its dust cover quotes from John H. Durston's foreword: "Koestler's experiences among the persecuted and displaced in this world have given him an understanding of human oddities that enlivens and enlarges his portrait of Kepler."[2]

Both Kepler and Koestler were interested in the nature of scientific discovery. Koestler shared Kepler's view that the road to discovery is hardly less noteworthy than the discovery itself. This approach is beneficial to later investigators for the road to discovery is often muddled or is outright lost. Koestler gives a succinct summation of his views about the nature of scientific discovery in the epilogue of *The Sleepwalkers*. To him, the synthesis of the new is not only an addition to the existing but also connecting what heretofore was without interaction. It is quite typical that the success of a new, revolutionary thought is preceded by some kind of overspecialization. The reintegration may happen within a specific area, but it may also happen between larger branches of science. Coming from the outside, especially if

bringing over considerable experience, maybe fertile for the new field. The disintegration of old dogmas, of course, helps give rise to new syntheses. Koestler's examples include Newton's new synthesis following Kepler's crushing the orthodox astronomy and Galileo's crushing the orthodox mechanics. The unification of modern physics and chemistry could be accomplished only after physics had given up the idea of indivisibility of the atom and chemistry of the impossibility of the transformation of the elements. Both those obsolete ideas used to be most progressive in their time.

In addition to the generalizations about discoveries, Koestler was seeking general features of the discoverers. He found them skeptical, irreverent toward authority and dogmas and having a tendency of questioning everything and anything without accepting it. He found them broad-minded to the fault that sometimes led them to be credulous. These traits facilitated them to be open to anything new and capable of looking at old things from a new angle. All these traits are components that contribute to the creativity of the discoverers. This is also why the questions the discoverers pose are often as revealing, sometimes even more revealing, than their answers to those questions.

It was not the factual material that I gained from reading Koestler that was useful for my physics lectures; rather, the atmosphere I soaked myself in as I was preparing for my class. The American student is very critical and is not easily dazzled by historical tidbits. When he or she feels that the lecturer is straying too far from the syllabus, the unease is apparent in the auditorium. It is almost an art to embed the additional material into the obviously mandatory core. My experience with my students gave me a strong indication that the effort was worthwhile.

---

[2] J. H. Durston, "Foreword," in A. Koestler, *The Watershed: A Biography of Johannes Kepler* (Garden City, NY: Doubleday & Co., 1960), p. 11.

Symmetries Revolting

Ferenc Lantos in 1982 in his workshop in Pécs (photograph by the author).

Ferenc Lantos (1929–2014) was an artist and a pedagogue in Pécs, south-western Hungary. He was Magdi's drawing teacher in general school, but my interaction with him developed independently. I had known his work before I took the British crystallographer, Alan L. Mackay, to visit him in 1982. The interest in symmetry was the link between us. Lantos prepared illustrations for some of our papers and books. In 1983, he asked me to write an Introduction to the catalog of his retrospective exhibition. I titled the piece "Symmetries revolting" and it was about the dangers of oversaturation by symmetry. Of course, it was not the symmetries that revolted in Lantos's creation, but the artist. He developed a modular approach to pattern design even if he did not use this term to describe it. To me his patterns created from the interference of other patterns represented special interest. They could be described as regular without being periodic. It was at about the time when Alan Mackay had predicted and then Danny Shechtman discovered the so-called quasicrystals for which it was characteristic to be regular, but non-periodic. Both Mackay and Shechtman figure elsewhere in this book.

© Springer Nature Switzerland AG 2020
77
I. Hargittai, *Mosaic of a Scientific Life*, https://doi.org/10.1007/978-3-030-34766-6_22

Lantos painting of both symmetric and anti-symmetric motifs (photograph by the author).

Lantos had numerous single-person shows and participated in group exhibitions. There was one single-person exhibition that has not been listed in any of his biographies. Our good friend, Alfred Lowrey, of the Naval Research Laboratory (NRL, Washington, DC) liked Lantos's paintings so much that he single handedly organized an exhibition in his home town, Greenbelt, Maryland. It was a most fortunate venue because Greenbelt is quite unique for its Bauhaus architecture—something most of its inhabitants are not aware of. Alfred and I met when I gave an invited lecture on symmetry in 1983 at the NRL. This was one of the highest-tension periods of the Cold War; suffice it to remember Ronald Reagan's "evil empire" speech. It was quite extraordinary that I received an invitation from the NRL—formally, it was from its association of scientists that organized general-purpose lectures. It was one of the largest audiences I have ever spoken to.

* * *

I had been familiar with the unique art of Roy Lichtenstein (1923–1997), and in 1994, one of his paintings caught my attention. It was "Peace through Chemistry" from 1970, consisting of three panels of which one was most expressly representing chemistry. It was at the time when I embarked on editing the new magazine about the culture of chemistry, *The Chemical Intelligencer*, so I decided to write a short piece about this painting. Chemistry has had a bad reputation, which has further deteriorated by the accusations of poisoning the environment. Here was a well-known artist who had something positive to say about chemistry. Besides, I liked Lichtenstein's geometrical style.

As it turned out, I misunderstood the message of the painting. I learned later that the artist's purpose was to convey irony rather than appreciation, and he wanted to warn of the possible misuse of chemistry—obviously, not for peaceful purposes. Fortunately, I was ignorant and judged the painting by the positive impact it had on me, and used it in the magazine. In fact, I used it also for the cover of the particular issue in which my piece was printed.

I still like the picture and am glad I trusted my instinct and ignored what might have been the original message. I also appreciate that the artist with this 1970 painting reached back to the Art Deco style of many American artists of the 1930s. It was the time of the Great Depression and the Art Deco style is still visible in the decoration of many public buildings in the USA. During the economic hardship, the federal government, Franklin D. Roosevelt's administration, supported the arts thereby saving artists from starvation.

The cover of *Culture of Chemistry* with one of the three panels of Roy Lichtenstein's "Peace through Chemistry".

Irony or not, Lichtenstein's painting helps rather than hurts chemistry through its warmth and quality. This is also why when Balazs and I made a selection of the best papers in the 6-year existence of *The Chemical Intelligencer*, 15 years after it had been shut down, we chose this Lichtenstein for the cover of the book.[1]

Lichtenstein was born in New York. He learned painting in summer school and attended an art college in Columbus, Ohio. In 1942, he studied technical subjects in the US Army and served in the European theater in World War II from 1943. After the war, he earned a college degree in Ohio and was employed as a graphic artist in Cleveland between 1951 and 1957. He was appointed to a university position in the State of New York and returned to New York City in 1963. From this point on, he became a recognized artist who was flooded with commissions.

* * *

---

[1] Balazs Hargittai and István Hargittai, Eds., *Culture of Chemistry: The Best Articles on the Human Side of 20th-Century Chemistry from the Archives of the Chemical Intelligencer* (Springer, 2015).

I was in my first year as a student of Lomonosov University in Moscow in the academic year 1961/62 when I met with the name of the Russian sculptor, Ernst Neizvestny (1925–2016). Nikita Khrushchev (1894–1971) was the supreme leader of the Soviet Union whose reign was more relaxed than Stalin's terror, and there was even a promise of further relaxation. There were, however, strong forces among the Soviet leadership that opposed the "thaw" as Khrushchev's policy was characterized. They found the uncovering of Stalin's crimes dangerous, and there was a chance of a return to old practices. Khrushchev must have felt the alienation of many of his colleagues in whose eyes he demonstrated weakness, almost capitulation. Under such circumstances, Khrushchev's visit to the exhibition of contemporary art gained out of proportion significance. He went to the Manezh Exhibition Hall in downtown Moscow in the company of the full Soviet leadership. In the section of abstract sculptures, Khrushchev demonstrated his rage in condemning what he saw in the most devastating terms. It was not known whether his unsophisticated taste revolted against modern art or just wanted to demonstrate the limits of his tolerance—perhaps, it was both. He made it clear that his judgment will have consequences for the Soviet art world and, probably, beyond. Under these most tense circumstances, a young, hardly known sculptor did the unthinkable: he contradicted the Soviet leader in the presence of everybody else and argued for the raison d'etre of modern art and for artistic freedom.

Neizvestny was the brave artist and it speaks in Khrushchev's favor that Neizvestny did not pay for his courage with his life—as he certainly would have under Stalin. He "only" paid with losing his livelihood, as he never again obtained commissions under the Soviet system in which artists lived only by the subsidies of the State. He received support from some leading physicists who recognized his difficult situation and who could afford—due to the strategic importance of their work—to stray off a little from the official order. It was not their only attempts to save real values under the dictatorship of Khrushchev and his successors. They represented the only, however slight, resistance to Trofim Lysenko's anti-science reign destroying the science of biology and sound agriculture. It was due to a commission from physicists that Neizvestny created the grave memorial for the Nobel laureate physicist Lev Landau in the Novodeviche Cemetery.

In an ironic twist, when Khrushchev, who had been deposed in 1964, died, his family wanted to choose the best Soviet sculptor to create his grave memorial, and they turned to Neizvestny. It now stands in the Novodeviche and masses of tourists visit it daily.

As Khrushchev died as a nonperson for Soviet officialdom, Neizvestny's creation did not improve the artist's standing in the eyes of the Soviet leadership. As a former supreme leader, Khrushchev might have been accorded a special burial place between the Lenin Mausoleum and the Kremlin Wall, but neither the leadership nor the family wanted that

Ernst Neizvestny and the author in 2014 in the artist's Manhattan studio-home (photograph by Magdolna Hargittai) and his Nikita Khrushchev's grave memorial at the Novodeviche Cemetery (photograph by István Hargittai).

resting place. After Khrushchev's memorial in the Novodeviche had been erected, the cemetery was closed for years under the pretext of renovation to avoid it to become a venue of pilgrimage of his admirers. When years later the cemetery was reopened and I went during a visit to Moscow, my camera was taken from me at the entrance and returned only upon my departure. When I complained about this at a friendly gathering in the evening, the next morning there was an envelope with a color slide of the grave slipped under the door of my hotel room—I never found out who the generous donor was. I needed the image as I wanted to use it in an article about symmetry. Neizvestny's sculpture over Khrushchev's grave radiated the dynamics of alternating black and white components and this made a magnificent representation of anti-symmetry.[2]

Eventually, the Soviets could no longer tolerate Neizvestny's presence and he was exiled. Magdi and I visited him in the fall of 2014 in his studio home in Manhattan. We made plans to visit his larger place on Long Island, but he died before this visit could have happened. Even his Manhattan base was huge, full with his art. He had just completed a book about his mother and he drew a sketch in the copy he gave us.

---

[2] Anti-symmetry is when a symmetry operation is accompanied by the reversal of a characteristic property.

Torvard C. Laurent and the author in 1996 in Stockholm (photograph by Magdolna Hargittai).

I came to know Torvard Laurent (1930–2009) through Lars Ernster (see a separate chapter) to whom I had been introduced by George Olah (also, in a separate chapter) in February 1996 at the University of Southern California. Laurent came to Budapest late spring in 1996 and this is when we met for the first time. However, his Hungarian interactions dated back to the start of his professional career. Laurent was born into a family of Swedish academics and was destined for a brilliant career in academia, but this did not make him conceited. He worked hard throughout his entire life and left nothing to chance. He was 17 when he embarked on his studies in medicine as a student at the Karolinska Institute in Stockholm. His interests focused on tissue cultures.

In addition to fulfilling the requirements of the curriculum, he started doing extra work in the laboratory and he was assigned to assist a recent immigrant from Hungary, Endre A. Balazs (see the László Bitó chapter). Balazs was an MD, who graduated recently from the Budapest medical school,

and his principal research was in polysaccharides. Balazs and Laurent started their joint work in 1949. Balazs was developing cell cultures under the strictest sterile conditions because there were yet no antibiotics helping their work. High molecular weight polysaccharides were needed for their research, and the umbilical cord was rich in them. These substances ensure the flexibility of the umbilical cord. Laurent's first task was to visit the delivery rooms in hospitals and collect umbilical cords from which they extracted the polysaccharides. Their cooperation lasted 15 months and resulted in three substantial papers. By the time Balazs left Sweden for America in 1950, Laurent was already an experienced researcher.

Balazs started his American career in Boston and converted a dilapidated laboratory into a well-functioning research establishment as part of the Retina Foundation. He continued his studies of polysaccharides. His research venue was connected with ophthalmology, the justification for the polysaccharide project was the high polysaccharide contents of the eye, in particular, the vitreous. In 1953, he invited

Laurent to join him as a visiting scientist. Laurent arrived and soon his fiancée, Ulla, as well and they married in Boston. Ulla was a practicing ophthalmologist. They stayed with Balazs until 1957 when they returned to Sweden but only for 2 years. From 1959, they were back with Balazs in Boston. Laurent was becoming so attached to his mentor that after two and a half years Laurent had to pose the question whether to stay with Balazs for good or return to Sweden right away. Laurent wanted to preserve his independence and he and his wife opted for continuing their life in Sweden.

The early 1960s was a good time to get established in Swedish academia. There was a tremendous expansion in science, huge investment in universities, the number of students grew rapidly, and there was almost limitless support for research. Laurent was appointed to the University of Uppsala and continued his research of polysaccharides. In 1963, he applied for the chair of medicine and physiology. There were 17 applicants, but the short list reduced only to two names, Torvard Laurent and Lars Ernster (see the Ernster chapter). It took two and a half years for the authorities to decide between them. Torvard was an MD and this decided it for him. The biochemist Ernster soon won an appointment to be Professor of Biochemistry at Stockholm University. Laurent and Ernster remained good friends and this was fortunate for me as well because three decades later their joint support provided the foundation for my exceptionally fruitful Swedish interactions.

Laurent had a brilliant career in Swedish scientific life. He enriched the chemistry of polysaccharides in physiology. As science is a factor in Swedish political life his career had relevance for Swedish politics. The listing of his functions is staggering: President of the Royal Swedish Academy of Sciences, Secretary (i.e., CEO) of the Wenner-Gren Foundation, Chairman of the Board of Directors of the Nobel Foundation. He was a long-time member of the Nobel Committee of Chemistry. He was awarded the most prestigious Swedish decorations. Being in charge of the Wenner-Gren Foundation, he presided over granting enormous sums of money in support of science. He remained a quiet, unassuming man throughout his life.

Laurent arranged for my invitation as a Wenner-Gren Lecturer in October 1996 on the day of the announcement of the physics and chemistry Nobel Prizes at the Royal Swedish Academy of Sciences. It was a trip to Stockholm from North Carolina where we arrived at the end of August to begin my visiting professorship at the University of North Carolina. In Stockholm, we attended the announcement of the chemistry prize, which was awarded for the discovery of the buckminsterfullerene, C_{60}, molecule. The connection between the award and my lecture about symmetry became obvious.

Fall 1996 was an unusual time. Shortly after we had arrived in Wilmington, North Carolina, hurricane Fran came. Magdi had left for Norway to be a referee of a doctoral dissertation, so I had to cope with securing our rented house. I covered all the windows with masking tape lest the pieces of glass fly away as rockets if the windows would break. Fran was arriving as predicted with the accuracy of a Swiss watch. In its eye, there was a whirling storm of an extreme speed of about 220 km/h (140 miles/h), but Fran itself was advancing at a snail's pace, which made the accurate forecast of its whereabouts possible. As soon as I had secured the house I drove away, deep inland—by evening I was about 300 miles away. Following landfall, I drove back the next day and the devastation astonished me. So did the discipline that the people of Wilmington displayed.

In 2000, Laurent and I organized a symmetry symposium in Stockholm. It was sponsored by the Wenner-Gren Foundation. The topic fit eminently the series of Wenner-Gren symposia that are often inter- and multidisciplinary. We had no limitations on inviting presenters from all over the world, all expenses paid, as long as they were top representatives of their respective fields. "Symmetry 2000" was the title of the meeting to demonstrate the state of the art in a diverse set of disciplines, including the arts.

The Hargittais (left) and the Shechtmans (right) in 2011 in Stockholm at the Royal Swedish Academy of Sciences during the Nobel Prize festivities (by unknown photographer).

At the opening ceremony of the Symmetry 2000 meeting, the Israeli Danny Shechtman received the Aminoff Prize for 2000. It is an award by the Royal Swedish Academy of Sciences for discoveries in crystallography. Shechtman received it for the discovery of quasicrystals.[1] In 2011, Shechtman received the chemistry Nobel Prize for the same discovery. (There will be more about Shechtman's discovery in the Alan Mackay chapter.)

Laurent and I edited a two-volume treatise of 52 contributions to the meeting.[2] I learned a great deal from Laurent during our joint work. He was meticulous in his editorial work, without any sign of fatigue, because his natural curiosity drove him in fulfilling his editorial tasks.

He was also instrumental for the publication of yet another of our books. In 1989, we published a book on symmetry for children in Hungarian (this book figures in the Lloyd Kahn chapter).[3] They printed 4000 copies of the Hungarian book and it sold out so quickly that in a few years' time we had to advertise in newspapers when we wanted to acquire a few copies. Laurent liked it and thought it as a great means for broadening education at an early age. The Wenner-Gren Foundation arranged for the translation into Swedish and the book appeared in Stockholm some 10 years following the Hungarian publication.

---

[1] Crystals have regular and periodic structures. In contrast, the structure of amorphous materials is not regular neither periodic. The quasicrystals, discovered by Shechtman in 1983, are regular, but non-periodic. At the time of their discovery the quasicrystals were considered to be outside of the realm of crystallography. By now, they have been incorporated into an expanded crystallography.

[2] I. Hargittai, T. C. Laurent, eds., *Symmetry 2000*, Parts 1 and 2 (London: Portland Press, 2002).

[3] Hargittai M., Hargittai I., *Fedezzük föl a szimmetriát!* (Budapest: Tankönyvkiadó, 1989). The Swedish version: *Upptäck symmetri!* (Stockholm: Natur och Kultur, 1998).

The Hargittais at the centennial Nobel Prize celebration in December 2001 in Stockholm (by unknown photographer).

In addition to my 1996 Wenner-Gren presentation, Laurent initiated another, at the Royal Swedish Academy of Sciences. This was a presentation about the Nobel Prize[4] on the occasion of the centennial of the Prize, on December 11, 2001, the day following the prize-awarding ceremonies. It sounds improbable that I was asked to give such a talk at this venue, but this is what happened. The title of my presentation was a quotation, "For many are called, but few are chosen" (*Matthew 22:14*).

---

[4] At this time, my book was already in production: I. Hargittai, *The Road to Stockholm: Nobel Prizes, Science, and Scientists* (Oxford University Press, 2002, 2003).

December 10, 2001, was the day of the prize-awarding ceremony in the Stockholm Concert Hall. It was part of the centennial celebrations of the Nobel Prize. On this occasion, in addition to the new laureates (first row, left) and the members of the Royal Family (first row, right), previous Nobel laureates are seen behind the first row on the stage. The Swedish academicians, who usually sit on the stage, were in the audience on this occasion. Our place was in the balcony among the press, which was fortunate as we could take pictures (it is prohibited from the audience). Photograph by Magdolna Hargittai.

Paul Lauterbur and Peter Mansfield

From Basic Research to MRI

Left: Paul Lauterbur in 2004 in his home in Urbana, Illinois (photograph by the author). Right: the author and Peter Mansfield in 2005 in front of the Sir Peter Mansfield Magnetic Resonance Centre of the University of Nottingham (photograph by Magdolna Hargittai).

A dozen years ago I was to have an MRI. I appeared at the clinic at the appointed time, but when I saw the tube I was supposed to climb into, I declined. When I looked at the technician apologetically, he told me that this happens, and I was not too special. Nonetheless, I felt a little embarrassed and regretted skipping the experience; the more so because I had met both scientists who had recently received the Nobel Prize in Physiology or Medicine for their contributions to the development of MRI.

The invention of *magnetic resonance imaging* (MRI) is one of the most beautiful stories in science history at more than one level. One is that the investigations had begun in pure science whose outcome then benefited millions. The other is that both laureates had to overcome considerable barriers in their respective careers before they arrived at the discoveries and their recognition. A discovery does not have to be directly beneficial to be recognized by a Nobel Prize; even a theoretical advance may bring its originator the Nobel recognition. In case of MRI, however, the direct benefit was enormous and long before the Nobel award, MRI had become a milestone achievement in medical diagnostics.

For MRI to happen a number of discoveries had to precede it. Going back to the beginnings, the existence of atoms had to be recognized, then, that the atoms have nuclei, and that they have magnetic properties, and, finally, that it is possible to communicate with those properties. The initial discoveries were of theoretical significance. Then, applications appeared, including nuclear weapons and energy production in nuclear power stations. The word "nuclear" has gained a frightening connotation hence it was dropped in the designation of imaging. This is how it became magnetic resonance imaging rather than nuclear magnetic resonance imaging.

Paul Lauterbur (1929–2007) and Peter Mansfield (1933–2017) received their shared Nobel Prize on December 10, 2003. Magdi and I were visiting our daughter Eszter in early 2004 in Evanston, Illinois, and on February 1, I drove to Urbana, Illinois, to visit Lauterbur. He was a fresh Nobel laureate, but the discoveries he received the prize for happened many years before.

When the technique of magnetic resonance had already become a widespread method in structural chemistry, the question arose whether it could be applied to solving

© Springer Nature Switzerland AG 2020
I. Hargittai, *Mosaic of a Scientific Life*, https://doi.org/10.1007/978-3-030-34766-6_24

problems in biology. The first step may have been when someone inserted a finger into the cavity of the sample to be examined. However, there were serious theoretical problems to be solved before any observation could be interpreted in a meaningful way. Seminal publications appeared in the early 1970s from a number of researchers, including Lauterbur of the University of Illinois at Urbana, and Mansfield of the University of Nottingham in England. Lauterbur's papers preceded Mansfield's by a few months, but this was not at all decisive as MRI was not a one-stroke discovery; rather, it was the result of series of discoveries.

Lauterbur was a dedicated researcher, so much so that he ignored writing a dissertation and becoming a PhD, he was so taken by his projects. Then, there was a few years delay in his career because of his military service. Eventually, he realized that becoming a PhD means a license for independence in research, so he decided to earn his doctorate. It often happens that in retrospect it is difficult to identify the determining steps on the way toward a discovery. In time, the emphases and relative significance change. Lauterbur told me that it used to upset him how animals were treated in scientific experiments. He called this a psychological trigger for his interest in developing MRI. However, this could not be too much on his mind when he described his first results because he stressed the theoretical advances and quite ignored the diagnostic possibilities. This is perhaps why the reviewers of *Nature* suggested rejection for his first manuscript in this project. By then, Lauterbur had become a recognized authority and he did not give up. He revised his manuscript and sent it back to *Nature*. The journal sent the revised version to a new reviewer. This reviewer still did not find it convincing, but he had known Lauterbur to be a reliable and trustworthy researcher and suggested publication. Lauterbur was 43 years old at the time and sufficiently experienced to recognize that if everything would work out as he expected, this might become something big. When he first realized the ideas that would eventually lead to the MRI, he jotted down his thoughts, had them signed by witnesses, and notarized the document by a notary public. Later, when he was unsuccessful in having his discoveries patented, this notarized document had decisive importance for establishing his priority.

Mansfield and Lauterbur worked independently of each other. Mansfield had daring ideas and deep theoretical foundation. His road to becoming a PhD was even longer than Lauterbur's and more difficult. It also demonstrated his exceptional dedication and will power. Before we visited him in Nottingham in January 2005, I read all that I could find about him. He left school at the age of 15, and his return to education appeared rather circuitous. At 15, he was a dropout, but it was not his failing; it was the system of education that failed him. He was among those London children who were sent into the country during World War II in order to protect them from the German bombing raids. His return to London at the age of 11 coincided with a decisive examination designated to decide the type of schools in which the pupils would continue their education. He went unprepared and passed the exam with average grades that qualified him for a middle level school whose graduates were not meant for a university education. This system soon changed, but this did not help Mansfield.

He was 15 when he completed his school and when he told the career advisor about his interest in space travel, he was laughed at and he was directed to study bookbinding. He eventually managed to switch to printing, which interested him more. He was a trainee compositor for 3 years. He introduced innovations and constructed a new small-size printing machine. From the start, he signed up for evening classes and took all the exams that he managed to enroll for. When he was 18, he read a newspaper article about a young man whose circumstances were similar to his and who managed to find a job in a ministry where he was involved in research. Mansfield contacted the editor of the newspaper, then the ministry, and landed a job that not only encouraged his involvement in research but also required him to keep studying. He succeeded in high school graduation, which was followed by 2 years of military service. He was 23 when he finally enrolled as a university student and various scholarships assisted his studies. He became a regular student and from this point on his career was on track.

Mansfield worked hard for his discovery and its success, and Lauterbur's pioneering activities facilitated the acceptance of his ideas. He was luckier than Lauterbur in that he could patent his innovations and he filed for a number of patents. Many years later, the income from patents made his early retirement possible. Thus, he could devote all his time and energy to the perfection of MRI. He founded a miniscule company and kept busy trying to solve two problems. One was to eliminate the claustrophobic feeling of the patients undergoing the procedure. The other was to diminish the disturbing noise during the examination. For years he felt close to the solution for both problems only to be disappointed again and again. He considered his work on these two problems more of a challenge than the invention of MRI.

The 2003 Nobel Prize in Physiology or Medicine went to these two scientists. According to the statutes of the Nobel Prize, there was a possibility to add one more person, but this option was not utilized. Raymond Damadian (1936–) protested with full-page advertisements in leading American and Swedish newspapers his omission from the Nobel recognition. He illustrated the ads with a copy of the Nobel Prize medal positioned upside down. The decisions about the Nobel Prizes cannot be changed; the Nobel Prize establishment is an independent institution. However, Damadian's claim was not without foundation. He had pioneering inventions, especially raising the initial question about the biological applications of magnetic resonance. Also, his inventions played a prominent role in making magnetic resonance apparatus available for medical purposes. He founded a company, produced instruments, filed for successful patents, and won patent trials. In an ideal world, all three slots of this particular Nobel Prize would have been utilized. It is also true that the motivation of this Nobel Prize, viz., "for their discoveries concerning magnetic resonance imaging" was not exclusionary. This motivation allows that others may have also had discoveries concerning MRI, except they were not included in the prize. The omission may be of one or multiple discoverers. We have no information on how the Swedish judges came to their decision and it may stay unknown forever. There are no minutes of the Nobel Prize deliberations. After the 50 years' time limit passes when the respective materials of the Nobel Prize archives become available for research, it is only the nominations, but not the deliberations that become available for scrutiny.

There have been other Nobel Prizes for which conspicuous absences generated controversies. I sought out the scientists who were omitted and our conversations enabled me to understand their situation and, in some cases, airing their grievances helped the unrecognized individuals. I was curious to learn more about Damadian and I made efforts to contact him, but they remained without response. It is quite possible that the expected controversy delayed the Nobel award for MRI. Richard Ernst lamented the delay in this recognition when, in 2002, he wrote that the disrespect for MRI in Stockholm is particularly difficult to understand. Ernst knew what he was talking about; he received the Nobel Prize in 1991 for the development of high resolution nuclear magnetic resonance spectroscopy.

Painful Anthem

Peter Lax in 2007 at the Courant Institute of New York University (photograph by the author).

The Abel Prize winner mathematician Peter D. Lax (1926–) was born in Budapest and immigrated to the USA with his family in 1941. During the 2000s Peter, his musician wife Lori, Magdi, and I met on a number of occasions, in their home on Central Park West, in our home in Budapest, and in our temporary home in Fort Lee, New Jersey. Lori was the daughter of the famous refugee mathematician Richard Courant. We attended memorable music evenings in their home. Peter's first wife was a mathematician, Anneli (née Cahn), who died many years before. They had two boys; the older, Johnny was a graduate student in history when killed in an automobile accident by a drunken driver. The younger, Jimmy, is a physician of internal medicine in Manhattan. Lax and I recorded a couple of conversations at the Courant Institute.

His father, Henrik Lax, was a sought after physician of internal medicine in Budapest. Peter's mother, Klára Kornfeld, was among the first women who graduated from medical school in Hungary. Lax attended the *Mintagimnázium* (Model High School) from which other celebrities had graduated before him, such as Theodore von Kármán, Michael Polanyi, and Edward Teller (there is a chapter about each of the two latter in this book). Lax excelled in math and in addition to the official curriculum received tutoring from two outstanding mathematicians. Rózsa Péter (1905–1977) was one; she taught in the Jewish high school and could get a job corresponding to her qualifications only after the war. The other was Dénes König (1884–1944), a pioneer of the theory of graphs, who

I. Hargittai, *Mosaic of a Scientific Life*, https://doi.org/10.1007/978-3-030-34766-6_25

committed suicide at the time of the Arrow Cross (the Hungarian Nazis) terror.

Lax did not graduate from his Budapest high school because his parents had the foresight and the opportunity to leave Hungary in time. However, he participated, unofficially, in an Eötvös Scholarly Competition for students who had already graduated from high school. He outperformed the official winners. In New York, he continued his studies in one of the top New York high schools, Stuyvesant HS. There, he was exempted from taking math, but represented Stuyvesant HS in math competitions with other high schools. Lax has a unique perspective to compare his Hungarian and American high schools, including the school atmosphere. His comparison is not void of his characteristic bittersweet humor. According to him, the American school prides itself that it prepares its students for life. However, on second thought, it may well be that the Hungarian school does this even better than its American counterpart. Lax excelled in his studies at the Model HS in Budapest, yet he, as all the students, felt paralyzed each time the teacher was calling somebody to the front for recitation. The students were afraid of their teachers. In the American school, the students considered the teachers their friends. In the Hungarian school, the student has to struggle for survival against the enemy—the teachers. In this, Lax saw the education for life.

Following Stuyvesant HS, Lax attended New York University (NYU) and became a PhD there. He spent his entire career at NYU. Richard Courant created a strong math school at NYU similar to the strong school he had created in Göttingen. This is the Courant Institute of NYU today and Lax has been one of its leading associates and for some time its director. In his youth, he spent his summers in California, at Stanford University, where two world-renowned mathematicians of Hungarian origin taught, George Pólya and Gábor Szegő. Mrs. Szegő and Lax's mother were cousins and Lax was staying with the Szegős. Their meals provided ample opportunity to discuss math.

Lax was conscripted at the age of 18 before he could have graduated from NYU. He received the basic training, then was subjected to an evaluation of intelligence, and was directed to continue his studies in subjects of technology and modern languages. He joined the Los Alamos nuclear laboratory in June 1945. There, he met Teller, but Lax no longer considered himself Hungarian. When he married in 1948, he did not want his wife to learn Hungarian. He could never overcome the repulsion he felt when reading about the horrors of 1944–1945. He had found out about that period from eyewitnesses, from books, and in particular from Béla Lévai's book, *Zsidósors Magyarországon* (Jewish Fate in Hungary). He is not accusing the present generations, but finds it an unresolved issue that Hungary still has not faced the past; rather, there is active falsification of history. As an example, he mentions the Terror Háza (House of Terror). It is

a sort of a museum remembering the Arrow Cross terror and the communist dictatorship, ignoring, however, 25 years of anti-Semitic discrimination and persecution, the *numerus clausus*, and the increasingly harsh anti-Jewish legislation of the Horthy era (1920–1944).

There is an ominous section in the Hungarian National Anthem, saying that

This people has suffered,
    For sins of the past and future!

The lyrics are from 1823. The poet may have not meant acquittal for crimes committed in the future though it sounds so.

Lax appreciates that Germany has faced the past. I would suggest though caution in considering Germany a model in this respect while recognizing that it has been considerably ahead of Hungary in facing the past. The denial of the Holocaust is a punishable crime in Germany, but should there be criminal procedure when something like what I am going to narrate here happens? In 2005, during a world science forum in Budapest I met a German international lawyer who represented the Lindau annual meeting of Nobel laureates at the Budapest Forum. At one point during our conversation, I mentioned that I was not yet 3 years old when I was put into a cattle carriage and the train was destined for Auschwitz. At this point, he interrupted saying that this cannot be so because 3-year-old children were not taken to concentration camps. He did not mean that I was not telling the truth; merely he thought the story too incredulous. I was upset and left him. After the event ended, he sought me out and apologized; of course, he knew that there were no age limitations for those who perished in Auschwitz. Had I not been his interlocutor though, he might have gotten away with such a falsehood, which I suppose was not premeditated, simply what had truly happened he found unacceptable. It should, of course, be unacceptable, except that it was acceptable in Nazi Germany.

In addition to his serving in Los Alamos in the summer of 1945, Lax returned to the weapons laboratory on later occasions. He was employed as a mathematician—rather than merely to carry out calculations as many less qualified associates were. His main task was the investigation of the explosive force of bombs of a variety of different shapes. His task was the mathematical description of shock waves that form as a consequence of the explosion. His results found applications in understanding the intricacies of supersonic flight.

Lax met all five "Martians" and he was most impressed by John von Neumann. According to Lax, there are not many that can see the principal directions of progress, and von Neumann was such a person. Also, there are even fewer that can adjust their own activities to be in concert with these principal directions, and von Neumann had this ability as well. This is why von Neumann who was originally theory oriented, was willing to devote so much of his time and energy to the development of the programmable computer.

Los Alamos gave Lax the benefit of recognizing the importance of computation. He was lucky to have spent most of his career in the Courant Institute, a world center of mathematics and computation. Of course, he greatly contributed to maintaining the leadership role of his Institute. Today, the geographical location of the scientist's activities has become less determining than it used to be, and speed has impacted the character of research. Communications happen with the speed of light. In his youth, Lax could be working on a problem for decades; leave it and return to it. By now, research has become more competitive and the speed with which discoveries happen has become a factor. This development has favored Lax because he was always engaged in more than one project. Of course, this is also a question of the researcher's temperament.

Road of Contradictions

Sándor and Olga Lengyel in fall 1966 during an outing of the Research Laboratory of Structural Chemistry (photograph by the author).

I met the name of Sándor Lengyel (1914–1990) for the first time when I received a notice from Eötvös University that my application to enroll as a student of chemistry was denied. For the reason, my poor performance at the entrance examination was given. The notice was signed by the Dean of the Faculty of Natural Sciences, Sándor Lengyel. This notice caught up with me in early August 1959, somewhere in the Mecsek Mountains where I was participating in a camping trip of the high school from which I had just graduated. The notice had a devastating effect on me. My first thought was, how will I be able to find my wife? I always imagined that my future wife will be another student. I was not completely inexperienced though in receiving such a notice. Four years before, my prior acceptance to the local high school was annulled at the last moment before the start of the academic year. Then, the reason was that my late maternal grandfather used to be a shop owner. This time, as I found out eventually, the reason

was that the second husband of my mother used to be a shop owner. To me, he was my stepfather, but we never went through an official adoption procedure.

The entrance examination could not be the real reason. The written part was routine; in it, it was not possible to shine or to fail spectacularly. I remember the oral examination clearly even after six decades. The chemistry part was, again, routine. One either knew the answers to the questions or not. Physics was exciting. The physics examiner asked me about my favorite area. I remember the feeling that it was not out of kindness, but in some way the prelude for trouble. Even the examiner was surprised when I named the mechanics of projectiles; perhaps, he could not have imagined anyone who liked it. I knew I could not fail in this and I did not. Still, I left the examination with uncertain feelings; something was not right. Years later, when almost all members of the examination committee became my colleagues, I asked

© Springer Nature Switzerland AG 2020
I. Hargittai, *Mosaic of a Scientific Life*, https://doi.org/10.1007/978-3-030-34766-6_26

two of them about my exam. Both remembered and both gave the same response. My rejection had nothing to do with my performance; they had their instructions and acted, accordingly.

Four years before, Mother fought for my acceptance for high school studies. Now, it was my task. This time an added difficulty was that it was hard to argue about my merits if I failed my entrance examination. However, every official who agreed to look into my case saw immediately the real reason of my getting the rejection (I was not allowed to see the papers about me). Finally, the Rector (President) of Eötvös University accepted me—he had a small discretionary quota. I was there among the freshmen and Dean Lengyel accepted each of us with a ceremonial handshake.

Our next meeting happened 6 years later, in 1965. By then Lengyel was the Director of the Research Laboratory of Structural Chemistry at the Hungarian Academy of Sciences. It was a new, but small institution, a good venue for launching my career in research. There was hardly any structural chemistry research going on yet when I joined the Laboratory and I was an ideal addition. Lengyel was charged with creating a research unit for structural chemistry; this was exactly my field. Lengyel spoke languages, excelled in math, and considered it his task to create the conditions for research rather than advance his own career. He did not put his name on the papers produced by the associates of the Laboratory that was customary elsewhere for a director or department head. We had one joint paper of theoretical studies, but for it our contributions were commensurable. I was independent from the start, which was rare then, and it is even today. He became a friendly colleague. The rejection of my university application never came up between us; probably he did not know about specific cases.

Lengyel was not a popular figure; he must have hurt many in his prior positions and not only in academia but also in the civil service after the war. Politics must have played an important role in his actions during the darkest period of the Soviet-type communist dictatorship in the early 1950s. According to an unpleasant episode, he publicly denied his brother, an internationally renowned laser physicist in the USA. Of course, at the time having an American brother could have caused the loss of any job. Many years later, this brother, Bela A. Lengyel (1910–2003), came for a visit, and Lengyel proudly introduced him around in the Laboratory.

Lengyel had a reserved personality, yet on rare occasions, I experienced his personal care. A couple of weeks after the invasion of Czechoslovakia by the Warsaw Pact countries in 1968, I felt his concern. The invasion was a decisive moment for it showed the impossibility of Soviet-style socialism "with a human face." Magdi and I were on vacation during the invasion and when I returned to the Laboratory, Lengyel asked me to his office. He told me that he had no idea how I felt about the invasion of Czechoslovakia and he did not want to know. But if I contemplated any protest, he strongly advised me against it. He stressed that this advice by no means should be interpreted as an expression of his opinion about the invasion.

In 2018, I learned about Lengyel's daring demeanor during World War II. His mother, his chemist sister, Piroska, and he hid Jews in their home at Belgrade Embankment in downtown Budapest. Utilizing their chemistry, they forged documents or changed existing ones by altering the designation of religion in them. In 1991, the two siblings were awarded the title "Righteous Among the Nations," the distinction given by Yad Vashem—the World Holocaust Remembrance Center—to non-Jews who risked their lives to save Jews during the Holocaust. In Lengyel's case, this was a posthumous recognition.

Internal Emigration

Alan Mackay in 1982 in Budapest
(photograph by the author).

I had known the name of the British crystallographer Alan L. Mackay (1926–) from the scientific literature years before we met in 1981 in person. It happened in Ottawa during the congress of the International Union of Crystallography. It was not a good beginning. I went up to him, introduced myself, we hardly exchanged a few words when he turned and left. In a few weeks' time, I received a gracious letter from him that he was glad we met and suggested that I visit him. Later I learned that our Ottawa encounter was typical of him; he was not for small talk and his hearing difficulties were already interfering with his interactions. He proved to be an open-minded, friendly person and we have become great friends visiting each other with our wives and staying in each other's homes in London and in Budapest. He already came to see us in September 1982, and we organized three lectures for him about symmetry. Two of the three were about fivefold symmetry, which is common in the world of molecules, but cannot exist in crystals according to the rules of classical crystallography. For this, there is a nice geometrical proof, which is often an examination question in related subjects, such as solid-state physics and materials science. The absence of fivefold symmetry in crystals has been a fundamental dogma, but it is characteristic of Mackay to question dogmas. He argues that if we accept dogmas uncritically, even if we encounter their violations, we may remain blind to them.

As it happened, already 5 months before Mackay talked to us in Budapest about the possibility of fivefold symmetry, an Israeli scientist, Dan Shechtman, made such an observation. As a visiting scientist at the National Institute of Standards and Technology (NIST), he was investigating aluminum/manganese alloys, looking for the ideal compositions for practical applications. One of the compositions of the alloy showed tenfold symmetry in his electron microscope/electron diffraction experiment. Tenfold symmetry was as much "forbidden" as fivefold.

Dan Shechtman (left) and Alan Mackay (right) in 1995 in the Hargittais' home in Budapest (photograph by the author).

Mackay did not know about Shechtman's observations, which reached publication only after more than 2 years; they were looked upon as utterly unbelievable. By the time of Shechtman's experiments though, Mackay had performed an exciting simulation experiment. To test the possibility of fivefold symmetry, he shone light onto a planar pattern of fivefold symmetry and detected the emerging diffraction pattern. It showed the forbidden symmetry and later it was found that it was consistent with Shechtman's observation.

Shechtman received the Nobel Prize in 2011 for his discovery. His road to recognition was no easy trip. The greatest chemist of his time, Linus Pauling, advocated an alternative explanation for Shechtman's finding, which remained within the domains of classical crystallography. Pauling who at his younger age himself brought down dogmas, now represented conservatism.

In contrast, Mackay, for whom it did not mean any trauma to throw out a dogma, did not find the discovery too extraordinary. He thought it was a consequence of the too restrictive definitions of classical crystallography. He thought that changing the rules would make Shechtman's observation lose the notion of novelty. He also enjoyed great authority in structural science, not as great as Pauling, but sufficiently great to make the Swedish judges pause when the question arose of awarding the quasicrystal discovery with a Nobel Prize. Mackay was too close to the discovery to appreciate its real value. In science, discoveries that broaden our world view are pivotal. Such was the discovery of quanta. Such was the development of the theory of relativity. When Mackay underappreciated Shechtman's discovery, he underappreciated his own contribution as well. When Shechtman's Nobel Prize was announced in 2011, nobody would have been surprised had Mackay been included, but he was not, and nobody protested his absence.

listened to what the authorities had to say, and kept their opinion to themselves. The inclination for internal emigration was strengthened over the years by his increasing hearing problem. He could always escape to his books; he was a voracious reader. He read books in English, but also in languages he did not speak.

Alan enrolled in Cambridge in 1944; the Nobel laureate W. Lawrence Bragg and the future Nobel laureate R. G. W. Norrish were among his professors. Alan was active in various student societies that expressed societal responsibility and he was a member of the student group who went to build a railway line in Yugoslavia after the war. Following graduation and a short detour in other jobs, he joined Birkbeck College of London University and stayed there for his entire career.

Left: J. Desmond Bernal around 1960 in London (courtesy of Alan Mackay). Right: Roger Penrose in 2000 in his office at Oxford University (photograph by the author).

Mackay was born in Wolverhampton, England. His parents were medical doctors and their dinner talk was always about their cases. His mother was active in social issues. Alan had to pass an examination when he was 8 years old in order to get accepted to a school his parents wanted him to attend. When World War II began, Alan helped the homeland front acting as a messenger. At 14, he was sent to boarding school. He had remarkable teachers who could not find better employment during the long years of economic depression and taught in school to the benefit of their pupils.

Alan had excellent grades, but already in his school years he withdrew into what he called internal emigration and stayed in it for his entire life. He just followed his ancestors who had developed an attitude according to which they

Alan worked as an associate of J. Desmond Bernal (1901–1971), but stayed independent. Still, Bernal had a strong influence on him in developing his universal interests. Several of Bernal's disciples would become Nobel laureates, but he himself did not carry through his original ideas. He was more interested in the societal impact of science than doing detailed research. Bernal was a communist and a great friend of the Soviet Union, and this sometimes put sad limitations on his objectivity. Alan followed Bernal in his broad interests, but never in his rabid Soviet bias. Alan's interest in languages made him learn Russian and there he met a fellow student, Sheila, who became his wife.

Fivefold symmetry was only one of Alan's many interests, but probably his activities in this area will have his longest-lasting impact. Rather than his individual discoveries, his contribution to our changing world view about the science of structures is what stands out in his activities. Over the years, he developed what is called generalized crystallography, which is not only about crystals but also about structures. From time to time he published reviews about generalized crystallography and it is instructive to read these papers in sequence. We can thus follow how his views and the views of the community, mostly under his influence, have progressed. He worked out something like a dictionary, which provides the modern terms and the corresponding ones of the classical teachings. His innovations came primarily from thinking, but he conducted experiments when he deemed them necessary, like it was the simulation experiment I mentioned above. Honoring him and his science, I organized twice special collections of papers from the international community in *Structural Chemistry*. I co-founded this journal 30 years ago and have continued editing it ever since. The appearance of the special issues coincided with Alan's anniversaries, but their real purpose was to take stock of where the science of structures stood at a given point in time.

Alan liked to view his work in a historical perspective and learned from such past greats as Albrecht Dürer and Johannes Kepler. He had a fruitful interaction with Roger Penrose (1931–) in devising new patterns and investigating their properties. Alan was among the firsts who recognized the importance of the so-called Penrose pattern and conducted simulation diffraction experiments on it. Alan's achievements were honored by his election to Fellow of the Royal Society (FRS), the distinction he valued most as it came from his fellow scientists. In 2010, he was awarded a share of the Oliver Buckley Prize by the American Physical Society. His co-recipients were Dov Levine and Paul Steinhardt who published a successful model shortly following the publication of Shechtman's discovery of quasicrystals.

Alan has been deeply interested in the nature of scientific discoveries. This was our principal common interest. His considerations extended from the most rudimentary to the most sophisticated. He liked to quote Bernal according to whom it was useful to be interested in many problems simultaneously because this enhanced the probability of a discovery. Of course, such things depend on the temperament of the scientist, but Alan's interests certainly had few limits. He was fascinated with the possible explanation of the conspicuous Jewish share in scientific discoveries. He thinks that Jews respect existing dogmas less than most others because often stupid laws and regulations served for suppressing them and discriminating against them. Bringing down dogmas may facilitate scientific discoveries as the example of the quasicrystal discovery demonstrated, but it was just one example among many.

Today, Alan, in his 90s, continues living in a sizzling intellectual atmosphere of his own making, as his wife of many decades has passed on and his connections to the outside world have further weakened with deteriorating hearing and vision. The air is still vibrating around him when he participates in conversation and his interlocutors face ceaseless intellectual challenges. He has collected a selection of his studies calling the volume pointedly *Eclectica*. For the scientific aspects, Alan is Bernal's true intellectual successor, alas, with Alan, this series breaks. We live in an era when instant results are expected and there is insufficient tolerance for a universal and inevitably skeptical approach to the big questions of science.

George Marx

Return to Fatherland

George Marx in 1999 in his office at Eötvös University (photograph by the author).

George (György) Marx (1927–2002) was a particle physicist; he popularized science and was a pedagogue. He made original discoveries, but his most consequential deed was his efforts in bringing back the émigré scientists into the collective mind of Hungarian society. These émigrés left Hungary to escape persecution. For quite some time they were treated as nonpersons by Hungarian officialdom, regardless whether it was Horthy's anti-Semitic autocracy, from which they had escaped, or the Soviet-type communist regime after World War II. They had become greats in science. For those that were still alive, Marx facilitated their visits to Hungary in the 1980s. For those that had already died, he worked on returning their ashes to be reburied in distinguished graves. This was the positive side of his activities. The negative was his efforts to mask the real reasons for the emigration of these scientists, which was, of course, consistent with governmental policy to avoid facing the past whether it was the crimes of the Horthy regime or Rákosi's ruthless communist dictatorship. How to handle the past was the issue where our views differed.

In fall 1999 I recorded a long conversation with Marx and although we had met frequently before, this conversation was what my sketch below is based on. Marx had a deep interest in history and in the role of physics in it. He was keen on establishing contact with great physicists, but it was only in the 1980s when his interest began focusing on the émigré Hungarian scientists. His first contacts were with Eugene P. Wigner (see a separate chapter about him) and John Kemeny.[1] Marx organized Wigner's visit to Hungary, he saw to it that Wigner had the proper conditions for a comfortable stay, recorded Wigner's talks, and had them published in the magazine of the Hungarian Physical Society, *Fizikai Szemle*. Marx was its Editor-in-Chief for decades and he shaped its character, making it popular not only among physicists but also a larger circle of intellectuals.

Marx was impressed by the émigrés, especially by Leo Szilard with whom he could not meet as Szilard had died long before Marx had embarked on his project. Marx was interested in what the Hungarian heritage meant for Szilard. It could not have been too much as Szilard was known to get upset when his interlocutors referred to his Hungarian background. However, Marx stressed the importance of the Hungarian experience. He said, "Cultures met and collided in Hungary in the twentieth century and the country was one

---

[1] John G. Kemeny (1926–1992) Hungarian-American mathematician and computer scientist.

of the focal points of history. Wars started here and political and ideological systems alternated here, in particular around the periods of the world wars. If a teenager who was open to the world experienced that the truths in the world of the grown-ups kept changing, that youth will not be devoted to some absolute truth; rather, his attention will turn to the changing trends in order to stay alive."[2] Marx often referred to Szilard that we need not be cleverer than others; we only need to take leave in time. Szilard practiced this when he left Nazi Germany one day before the borders were closed.

Marx boasted the large number of Hungarian Nobel laureates that was unreal on two counts. First, there were not so many Hungarian Nobel laureates that the Hungarian officials stated, stamps showed, and memorial plaques of marble commemorated. Second, there was never a hint about the persecution these laureates suffered before their emigration and that they were chased out of Hungary. Only after they had earned their fame, and in most cases after their death, Hungary had accepted them back and basked in their glory.

When confronted with the truth, Marx's defense was that he wanted to encourage greater support for science by exaggerating reality and wanted to lift the country's international prestige. The first claim was problematic because if science is flourishing even under the conditions of very low support, why should the government, any government, provide greater support? The second claim was questionable too. Even a superficial examination of the numbers of Nobel laureates in different countries leads to a pattern conspicuously different from Marx's claim. If the Austrians would count their Nobel laureates the same way Hungarians do, Austria would win such a competition by a large margin.

In absolute numbers, the USA leads the pack by far. Many ascribe this success to the importation of scientists. There is some truth in this. The influx of foreign scientists is facilitated by the better conditions for research and especially by conditions that enable younger scientists to become independent much sooner than in most other places, including Western Europe, let alone Eastern Europe, and the rest of the world. This is why not only Hungarian, Polish, or Chinese scientists have immigrated to America and become Nobel laureates there but also Belgians, Germans, and other West Europeans as well.

However, ascribing the large number of American Nobel laureates primarily to foreign scientists is misleading. There are many American high schools that have graduated future Nobel laureates. There are at least 13 high schools in New York City that have had at least one future Nobel laureate. Quite a few had more than one, and one even has had seven(!).

In our private conversation, Marx viewed the situation more realistically than in his papers and public statements. He told me that in Hungary the encouragement of school children and students for higher performance comes primarily from their parents rather than from their schools. The society as a whole is unable to appreciate even the most creative minds, and by appreciation he did not mean financial reward; rather, paying attention to them. The gifted young people may have found their environment sufficiently challenging, but the appreciation has been lacking, something that they found elsewhere where they had moved. For Marx the large number of Jews among the Hungarian Nobel laureates was a puzzle that needed investigation. After the 1867 Compromise between Hungary and the Habsburgs, Hungary became a tolerant land and an attractive target for Jewish immigration. The Jews played a conspicuous role in the modernization of the country between 1867 and 1914, the start of World War I. The loyal Jewish community enhanced the Hungarian population in the multiethnic Hungary in which the Hungarian population was in the minority. Following the so-called Trianon Treaty, when the country shrunk considerably, the Jews were no longer needed to enhance the Hungarian population. Moreover, anti-Jewish discrimination, exclusion, and persecution had a strong economic motivation—the expropriation of Jewish property.

The periods of crises in Hungarian history of the twentieth century meant extreme hardship for the Jews—Marx continued. He showed in charts that the émigré scientists so successful in the West had gone through the most difficult years of their lives in the war periods in Hungary. Apparently these trials helped them to develop their creativity. I should better quote Marx verbatim: "In such historical situations pregnant with tragedy, the values advocated by the grown-ups lose their credibility, and the correct anticipation of the next political regime, the next dominating ideology, becomes literally vital. The question to decide is to accommodate the new regime or to escape. I am convinced that the past Jewish experience in alternations of history facilitated their correct evaluation of the consequences of these rapid changes. ... For a teenager, it is hard to survive the rapid succession of vastly different historical periods, yet it is also good schooling. In this, I see the advantage of Jews in such highly perilous historical periods." I found it rather peculiar to speak about advantage in such context. Marx tried to justify his way of treating the past in his public appearances. He said he found it impossible to emphasize the Jewish roots of the scientists he talked about. If he identified any great physicist as a Jew, his anti-Semitic colleagues would attack him immediately. He added this diplomatic sentence: "Hungarian society is not extremely intolerant, but for a considerable segment of the society, including the intellectuals, the label of Jew is not indifferent."

---

[2] Hargittai István, "Beszélgetés Marx Györggyel." (Conversation with George Marx) *Magyar Tudomány* 2003, 883–889. My translations.

There may be something in what Marx said. To this day, mentioning someone's being Jewish or being of Jewish origin is considered impolite in Hungary. Of course, such a mention cannot be considered cut off from context. It may be condemnation and it may even be appreciation although even in that case something is suspicious because merely being Jewish is not a merit. The story below may or may not be an illustration of what Marx might have meant.

Edward Teller's election to the Hungarian Academy of Sciences came up in 1990, right after the political changes, when the one-party system was giving way to a multiparty democratic system. Here I narrate the story on the basis of published reference and archival material in the Teller Folder at the Archives of the Hungarian Academy of Sciences.[3] A memorandum was prepared in February 1990 for Teller's election to honorary membership of the Hungarian Academy of Sciences. In this memorandum, there are four references to Teller's being Jewish. One mentions that one of Teller's high school teachers used to address the class as "Gentlemen and the Jews!" Another is that Teller's father did not see the conditions in Hungary encouraging for his Jewish son's scientific career. The third mentions that Teller's family had to wear the yellow star under Nicholas Horthy, was incarcerated in a ghetto under the Hungarian Nazi leader Ferenc Szálasi, and was exiled from Budapest to the countryside in the early 1950s under the communist leader Mátyás Rákosi. The fourth reference is that Teller always stressed both his Hungarian and Jewish origins. The final, "official" nomination followed the memorandum closely except that it did not contain any of these four references to Teller's being Jewish. Teller was elected honorary member in May 1990.

---

[3] There is a two-page document of February 15, 1990, in the Archives of the Hungarian Academy of Sciences. This paragraph is taken largely from Balazs Hargittai and István Hargittai, *Wisdom of the Martians of Science*: *In Their Own Words with Commentaries* (World Scientific, 2016), p. 160.

She Was a Pioneer

Barbara Mez-Starck in 1999 (courtesy of Natalja Vogt).

As described in previous chapters, upon my graduation from Lomonosov University in 1965, I returned to Budapest and started my research work. For molecular structure determination by electron diffraction, I had to create my experiments out of nothing. We did not copy anybody, but produced new approaches in our quest for the best solutions. This was a very small area of science, certainly not a fashionable one, in which it was easier to excel. We became visible internationally, at least to a small circle of scientists, by the early 1970s.

The most telling sign of recognition was that visitors started coming to us and not only for a few days but also for working with us for months and even an entire year. A professor of the Norwegian University of Technology, Jon Brunvoll, spent a whole year with us on two separate occasions. Aldo Domenicano of the University of Rome, later, of the University of L'Aquila, came several times, each time for three-month stays. Others from Norway and from Italy, also from the USA and elsewhere came, but the most fruitful interactions with dozens and dozens of joint papers developed with Norwegian and Italian colleagues.

Barbara Starck, later Mez-Starck (1924–2001), from Germany, West Germany as it was then, initiated cooperation with us in the mid-1970s, and it was something special. In the 1960s Barbara was an assistant of a well-known professor at the University of Freiburg. The professor charged her with collecting the data he wanted to use in his lectures. She created an easy to handle database that other professors were also free to use and they did. Upon the sudden death of Barbara's professor, the decision was made that she should continue building the database as her main job. The task was restricted to gas-phase information, but it was inclusive for all data resulting from any physical, theoretical, and computational techniques. She was not an expert in all fields, but created a network of experts. This is how and why she contacted me. For years I stayed involved in the work concerning all structural data by our technique, gas-phase electron diffraction.

This was still the pre-Internet era and she arranged for us to receive with regularity all the necessary information on paper about all the publications in our field from all over the world. This was a godsend for our staying informed and my task was merely the critical evaluation of the studies. Even when the project became a full-fledged data center, Barbara continued her hands-on involvement.

From early on, she was concerned about the continuation of the project although she was far from retirement. The financial side of maintaining the center did not bother her. Recently, she had come to a substantial inheritance and she had no other ambitions. She was ready and willing to use her own resources for letting the center carry on under any

I. Hargittai, *Mosaic of a Scientific Life*, https://doi.org/10.1007/978-3-030-34766-6_29

circumstance. Problems might be anticipated elsewhere, not only in finances. Any unit at a university has to justify that its operation is needed for teaching and research. If the financial resources are missing the project dies. However, the existence of a project cannot be guaranteed just because the necessary funds are available.

Barbara had tremendous foresight. She concluded that the center needed a person to be in charge that would consider it his or her mission to carry on with the gathering, evaluation, and dissemination of the structural data. She talked with me about this at great length; I was a sounding board for her as, obviously, I was not an interested party—obviously for I had different ambitions. Of course, the person she was looking for had to satisfy some fundamental requirements, had to be a PhD, and had to have some prior experience in scientific research. Beyond that, she was looking for somebody who would consider it to be the task to develop and maintain the center at its existing goals—we might say the person had to have *limited ambitions*. Ambitions at a university setting mean devoting ever increasing energy and time to one's own research, becoming a professor, creating one's own research group, teaching one's own subject and suchlike. Such ambitions could be satisfied only at the expense of the center. Once Barbara formulated the "limited ambitions" concept, she felt invigorated in her search for her successor. In this, also, she was successful. Already, from the perspective of several decades, we have witnessed that her center has carried on without changing its original mission. It does not mean ignoring technical innovations and this is happening on a permanent basis. What has not changed is the concept of the center.

The cooperation with Barbara was an intellectual challenge. I was not interested in her decisions concerning personnel, but participated actively in creating and developing a whole new project. Then, this active and personal participation came to an abrupt end in 1983. Barbara had a stroke, the left side of her body became paralyzed, and although she carried on for years, everything cost her tremendous pain and disproportionally huge effort. Fortunately, her successor had had sufficient time to learn the trade and continued reliably. For Barbara, the next two decades were torture, but she could also witness the survival of her creation. To the end, she continued gathering data. For the last 15 years, she worked for free, but this did not bother her; she was wealthy. She spent as much money as she needed for her treatment; it did not shake the endowment of her center. She secured its future in the form of a foundation. She was always the person of action, and no hardship could prevent her from reaching her goals. Hardships had shaped her life and personality.

It was only a few years in her childhood that she could live the carefree life her parents meant to provide for her. It was a romantic story of how her German industrial magnate father, Hermann C. Starck, and her Hungarian-Jewish mother, Klára Sarkadi, got married. Klára was one of Starck's many employees. He had rigorous rules that precluded any close relationship between employer and employee. When one day Starck fired Klára, she was devastated, because she had always worked to full satisfaction of her superiors. This was changed into a great surprise when in the next moment, he proposed to her. After Barbara, there was one more child, Barbara's younger brother, Gerhard.

The idyllic life of the family changed when the Nazis came to power in 1933. The mother survived by hiding and bribes. The father spent a fortune to ensure as close to a normal life for the children, and as long, as it was possible. When Barbara graduated from high school she started her studies in chemistry at Freiburg University, but she had to quit in 1942, after one semester. She found employment in a plant working with metals and she did analyses in a chemical laboratory. All four of the family survived, but all Barbara's maternal relatives perished in Auschwitz.

After the war, she continued her studies, earned her Diploma (master's degree equivalent), and then in 1959 her doctorate. She stayed at the university and gradually developed the activities I described above. Barbara's mother died a few years after the war. The father became alienated from his children, remarried, but kept close control over Barbara and Gerhard and they were financially dependent on him. He disapproved the man Barbara was living with and she could marry only after her father died, when she was 50 years old.

Barbara had a reserved personality. We hardly talked about her background; still I felt her affinity toward "Eastern European" life. She behaved natural about her Jewish roots; did not hide them, but did not stress them either. Our friendship was based on our professional interests and cooperation. I do not think she saw her center for structure documentation as her memorial, but in fact, it was—it is. Today it is called the Chemical Information Systems at the University of Ulm and I expect it one day to be named after Barbara Starck. She had no such ambitions; it was as if she would like to remove her own self from what she had created though all her energies and resources were put into that institution. She willed to have all her personal affairs to be destroyed.

Mutual Interests

Kurt Mislow and the author in 1997 at Princeton University (photograph by Magdolna Hargittai).

Considering the time of the beginning of our interactions, the phrase "mutual interests" sounds pretentious. By the time I had developed my interest in symmetry in the early 1980s, Kurt Mislow (1923–2017) was already a world-renowned professor at Princeton University. He published a small book *Introduction to Stereochemistry* in 1965 in which he surveyed the chemical significance of symmetry and chirality.[1] I valued greatly this book, which gave me the idea of asking him to review a chapter of our symmetry book in preparation.

Looking back, I need to say a few words about the story of our first symmetry book.

When in 1978, my brother, Sándor, seven and half years my senior, defected from Hungary and moved to Israel, the air froze around me, the number of my friends diminished, and my passport was taken from me. Suddenly, I had some extra time and I embarked on writing a small book on symmetry. It was later developed into a book in English with the same title, *Symmetry through the Eyes of a Chemist*. The Hungarian book appeared in 1983; even for a small book it took the publisher 5 years to bring it out, but this was not exceptional in those days. The book had a novel approach to survey the presence and significance of symmetry in chemistry. It was well received and it was the base on which Magdi and I, now jointly, produced the first version in English of a more comprehensive book. We did this during my visiting

---

[1] Chirality refers to the property of our two hands that they are each other's mirror images yet they cannot be superimposed to correspond to each other. There are countless molecules that have "left-handed" and "right-handed" versions. The physiological effects of the two versions may be vastly different hence the pharmacological importance of molecular handedness or chirality.

© Springer Nature Switzerland AG 2020

I. Hargittai, *Mosaic of a Scientific Life*, https://doi.org/10.1007/978-3-030-34766-6_30

professorship at the University of Connecticut (1983–1985). This new book had much more mathematics in it than its predecessor, and this was due to Magdi's efforts. I am stressing this because people usually think the opposite, yet it was Magdi who made the book mathematically comprehensive. This book appeared in 1986, the second edition in 1995, and the third in 2009. Some schools have used it as a textbook or auxiliary text. From the first edition, it was a success and positive reviews appeared about it in the most authoritative venues.

It has been our working style to ask friends and colleagues, sometimes even people we had never met to review our manuscripts. For the *Symmetry through* book, it was not difficult to find reviewers for the more "chemical" chapters as well as for the more mathematical ones. However, this was not the case of the crucial second chapter, which discussed simple and more complex symmetries and went much beyond chemistry. We asked the famous Kurt Mislow to review this chapter for us and he kindly agreed. At that time his kindness might not have felt so extraordinary; today I understand better, how much it meant. He was at the peak of his activities, with a large group and research program, and teaching obligations. His comments arrived within a reasonable time frame. They were detailed and constructive, providing a tremendous help for us in improving our manuscript. I like to think he must have enjoyed doing this; otherwise it would have been impossible for him to spend so much time on the manuscript of these two unknown colleagues.

Subsequently, we met with Mislow many times, first in Darmstadt, Germany, at a symmetry conference. Then, for a period of several years, we used to visit Princeton annually as our daughter, Eszter, did her PhD studies in sociology at Princeton University. When we were engaged with our big interviews project, I recorded a long conversation with Mislow. We talked about many things. I was very interested in his chemistry, and he narrated his adventures unhurriedly. We talked about his life and this was one of those recordings when I think we both forgot that it was a recording for the purpose of eventually getting it published. It was a fine conversation about a host of topics of mutual interest. It was my standard approach to personally transcribe the text and send it to the interviewee for review. I did not want to publish anything without the opportunity to correct and even change what had been said. Soon I received back the text from Mislow, with hardly any change, but with a surprising note. It said that he was reading my transcripts with increasing astonishment. He found that he told me about things in his life that he had forgotten or thought he had, and about things that he had never told even his wife before. But he left everything intact.

Kurt Mislow was born in Berlin into an upper-middle-class Jewish family. His father was a businessman and they moved around a great deal. Kurt grew up in Düsseldorf. He witnessed bloody clashes in the streets before Hitler's accession to power, between the communists and national socialists. After 1933, the communists had disappeared, but the street disturbances continued and Kurt witnessed the humiliation and persecution of Jews. One of the most horrifying events was the widespread and violent pogrom on the night of November 9 to 10, 1938, the so-called *Kristallnacht* (crystal night, or, the "Night of the Broken Glass"). By then, the Mislow family was no longer in Germany. They moved to Italy in 1936.

Kurt kept his German passport, which has an image of the swastika between the paws of the German eagle and it is covered by a big, red letter "J" to warn anybody who looked at this passport that its bearer was a Jew. Kurt used to think that it was a German initiative to stamp Jewish passports with this discriminating sign, but this was not so. The Germans introduced this system at the request of the head of the Swiss Federal Police, Heinrich Rothmund, who thought this would facilitate spotting would-be refugees and preventing them from crossing into Switzerland.

The Mislow family lived in Milan from 1936 and Kurt attended a classical gymnasium. He had to learn Italian right away because he had to study Greek and Latin and the instructions were in Italian. He was happy in Italy; the anti-Jewish laws in Italy had yet to come. Until Mussolini had become Hitler's close ally, his fascism did not bother the Jews. Primo Levi, the famous Italian author writes about the difference between the post-World War I Italy and Hungary. He learned from his father who had spent some years in the 1920s in Hungary about the fear in which the Hungarian Jews lived under Horthy's anti-Semitic regime. His fears stayed with him for years even after he had left Hungary. Italian fascism stayed free of anti-Semitism for a long time. Levi talked about this experience in the 1980s in his conversations with the writer/journalist Ferdinando Camon.[2]

In 1938, Kurt moved to England and studied in a boarding school with all its horrors. But it was in this school where he had serious physics and mathematics for the first time. In his graduating year, he found a chemistry book and it led him to a life-long commitment to chemistry. When World War II broke out, the Mislow family was declared enemy aliens. Having acquired the necessary papers, in 1940 they moved to the USA. Kurt was a library helper first, but soon Tulane University of New Orleans offered him a scholarship and there he earned his bachelor's degree in 1944. Next, he enrolled for graduate studies at the California Institute of Technology (Caltech) and he became a PhD in 1947 under Linus Pauling's mentorship.

---

[2] Ferdinando Camon, *Conversations with Primo Levi* (translated into English by John Shepley, Marlboro, Vermont: The Marlboro Press, 1989), pp. 5–6.

Kurt witnessed racial discrimination in New Orleans, which stunned him for they had to leave Germany where he was the victim of discrimination himself. He traveled the tram every day and it was segregated. A movable metallic plate on the seat backs said "for whites only" and when all the seats for whites were taken the conductor moved the shield further back so that more seats for whites could become available. He encountered anti-Semitism as well. The local chapter of the honor society of chemistry students, Alpha Chi Sigma, did not take Jewish students, so Kurt did not become a member. It was for white Christian men only. Kurt was curious about when did they change this practice. Jewish chemists became eligible in 1948; black chemist men in 1954; and women chemists in 1970.

At Caltech, Kurt attended Pauling's beautiful lectures in quantum mechanics; he found it an uplifting experience. After he had become a PhD, Kurt got a job at New York University (NYU) from where he then moved to Princeton for a brilliant career. He became Emeritus in 1988, but only in an administrative sense. He continued his research with young associates, enjoying the support by the National Science Foundation.

He was probably the best-qualified scientist in the world who could help us in perfecting our second chapter of *Symmetry through the Eyes of a Chemist*. He was interested in the broadest aspects of symmetry. He could be considered to be the successor of both Immanuel Kant (1724–1804) and Louis Pasteur (1822–1895). Kant was deeply interested in chirality and thought it to be a paradoxical phenomenon. There is a paradox in that while in an ideal case, the left hand and the right hand are isometric, i.e., have the same metric properties, they are not superimposable, that is, they cannot be made to coincide spatially. Kant used the expression "incongruente Gegenstücke," non-superposable counterparts for such pairs. Lord Kelvin applied for the first time the terms chiral and chirality. By now, the issue had become much more than merely of theoretical interest. Chiral purity has become one of the cornerstones of pharmaceutical considerations and the production of medicine. Sometimes one reads that Mislow had introduced chirality into chemistry, which is an obvious exaggeration and Mislow was the first to protest it. It was Louis Pasteur who demonstrated in his experiments the connection between macroscopic chirality, as it appears in crystal shapes, and molecular chirality.

Mislow followed the success of our *Symmetry through the Eyes of a Chemist* with sympathy and support. During our last meeting, we talked about a joint booklet that would provide practical exercises to facilitate the use of *Symmetry through* for instruction. Mislow's rich, decades-long pedagogical experience would have made such a project a sure winner. Sadly, it was not to happen.

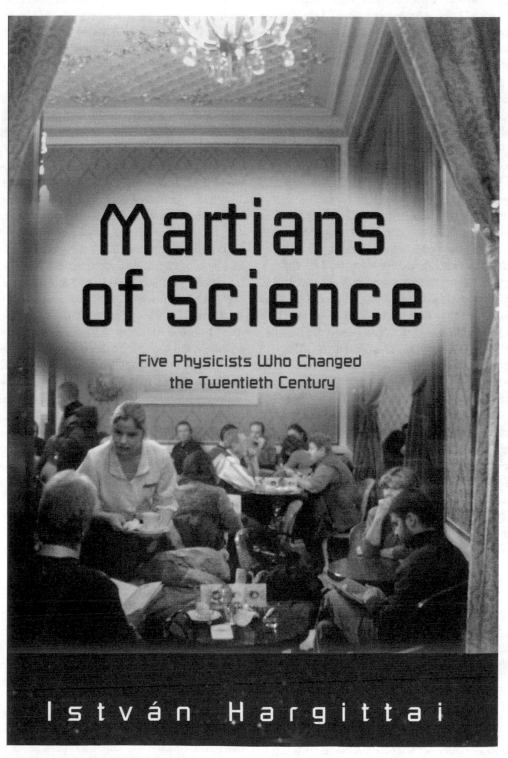

The soft cover version of the Martians.

© Springer Nature Switzerland AG 2020

I. Hargittai, *Mosaic of a Scientific Life*, https://doi.org/10.1007/978-3-030-34766-6_31

My book, *The Martians of Science* was first published in 2006.[1] The cover of the original edition, showing the portraits of the five Martians, was rather pedestrian. There was interest in the book, but nothing special. There were a few welcoming reviews and on the whole I was happy. University publishers are not spoiled by loud successes, so I hoped my publisher was also happy. I had no direct contact with the publisher because my editor had left the company before the book came out. I could not even thank him for his initiative. This happened with me for the first time that I had been asked to write a book. When in late 2003 he first suggested to produce it about this peculiar group of Hungarian-American scientists, known collectively as the Martians, I declined. It was quite at an early stage of my involvement with book publishing and I did not yet understand the significance of such an invitation. Besides, I did not think there was much that had not been covered about them in previous publications. I wrote back that there were better experts of the Martians in Hungary than I. His response was that he was not looking for the best expert but for a good author. I published a book in 2002 with Oxford University Press: *The Road to Stockholm* about the Nobel Prize and Nobel laureates and it was a success.

The five Martians, Theodore von Kármán, Leo Szilard, Eugene P. Wigner, John von Neumann, and Edward Teller, were not only distinguished scientists. They risked their scientific careers when they decided to devote their activities to the defense of the USA and the Free World, first against Nazi Germany and later against the imperialistic efforts of Stalin's Soviet Union. The label of "Martian" originated from a widespread story during the Manhattan Project. The Hungarian participation was conspicuous in the Project and Enrico Fermi posed the question about how it could have happened. Szilard's response was that in reality they had come from the planet Mars and spoke Hungarian to camouflage their origin. This anecdote has been around in several versions. The five Martians all came from upper-middle-class Jewish families in Budapest. They attended excellent high schools—the five of them three different ones—and fled the anti-Semitic Horthy regime following World War I. They went to the then democratic Germany and after Hitler took over Germany, they moved to the USA.

I began my work on the Martians book by gathering and reading everything by and about them, after having defined who exactly belonged to this group. This was my program for the first year. The second year was for writing. When at a book launching, someone asked whether there had been any moments of frustration during my work, I responded with

true frankness that there were none. Magdi had to make an effort not to contradict me as she did remember periods of high tension during which I was so absorbed with my project that I had to disappear from sight. Of course, as soon as Magdi reminded me of this, it all came back; otherwise I tended to erase any past frustration from my memory. The principal source of tension was that until I produced my first draft, I could not be sure whether there was indeed a genuine justification for a new book about the Martians. When the first draft emerged, I knew there was plenty.

The book was published mid-2006 and nothing special happened until spring 2007 when suddenly its sales jumped and the book became unavailable. I had no editor to inquire from about what happened and the person in charge of marketing showed ignorance not only in my project but also as it later turned out in everything where marketing should have been in action. Later I understood that as a rule when the number of available copies would sink under 150, they would reprint the book. This was a safe approach because such books as mine—in popular science—usually do not sell rapidly. *The Martians of Science*, however, was selling so fast in spring 2007 that even when there were still hundreds of copies its shortage could have been anticipated. Eventually I learned the origin of the sudden popularity of the book. The well-known businessman and investor, Charles T. Munger (1924–), better known as Charlie Munger, named a book annually that he liked and recommended to others. In the spring 2007 meeting of the shareholders of his company, he named my book and told the gathering: "A hell of a book about five Hungarian physicists driven to the US by Hitler and who contributed much to science here. I can't recommend it enough." Soon enough the Publisher decided to speed up the second printing and they did not let the book go out of print anymore. In 2008, they brought out a soft cover version with a more interesting cover (see above). With Munger's permission, his statement about the book was reproduced on the back cover.

I have been both lucky and unlucky with my book editors. Lucky, because I met dedicated editors who liked the projects we were doing together, and I learned a great deal from them. Unlucky, because there was no editor with whom I could have worked on more than two projects, mostly it was only one. They retired, became terminally ill, or changed their employment and could not take my project with them. All my American publishers were university publishers with one exception, Prometheus. There, Linda Regan had wanted to work on a Teller biography for a long time and when I came up with such a proposal, she thought it was a good match. My Teller biography (see the Teller chapter) was a consequence of the Martians' book. Working on it made me recognize the need for a balanced approach to present Edward Teller, admired by many and hated by, perhaps, even more. Linda was an old-fashioned editor; she needed all submissions on

---

[1] I. Hargittai, *The Martians of Science: Five Physicists Who Changed the Twentieth Century* (New York: Oxford University Press, 2006). It appeared in a soft cover version in 2008 in which the definite article in the title was omitted. The book has been kept in print ever since.

paper; read everything sentence by sentence and made critical remarks copiously. She wanted to prune some sections and expand others; she was never satisfied having some scientific term without an easy to perceive explanation. Once I returned the revised manuscript to her, the process started all over, but after the second revision, she was happy. I have always been eager to improve my composition and learned a great deal from Linda.

The first person I learned from about writing was my teacher of the Hungarian language in the first year of high school 1955/56 (see the Mihály Csonkás chapter). Then, in 1969, during my stay at the University of Texas at Austin, I signed up for a one-semester course on how to write a paragraph. For 12 weeks, all the weekly lectures and exercises were about how to compose a paragraph—it was very useful. Writing is an organic part of the research scientist's activities even if the scientific papers are rather dull and use a limited vocabulary. I have noticed that the papers of some authors are more interesting and not necessarily for a more interesting scientific content. I found some authors whose style impressed me and there were four whom I considered my remote teachers in writing technical papers. English is a second language for me, so I had to make much extra effort. How lucky the native English-speaking students are. It is not only that they own the language to start with but they also have to take English in their college curriculum—making their situation doubly advantageous. Such requirements include all students, including those planning to major in the sciences and technologies. The training of efficient communication cannot be started too early. Our American grandchildren, as young as 8 and 10 years old, had already prepared poster presentations on topics they had researched. Returning to Linda Regan, there was a second book for which she acted as editor, but fell ill before it could have appeared, resigned from her position, and soon passed on.

My indirect experience with Charlie Munger was a reminder of the benefits of not letting ourselves to be narrowly classified; rather, trying to broaden our horizons. Someone asked me whether I had moved into the financial world. When I looked puzzled this person told me about the appearance of information about my *Martians* book on a site for investors. It was not that I had moved anywhere; rather, it was the sign of the broad interests of a world-famous investor.

Soldier, Scientist, Politician

Yuval Ne'eman and the author in 2003 in the Hargittais' home in Budapest (photograph by Magdolna Hargittai).

My first visit to Israel was in 1992 at the Ben-Gurion University of the Negev in Beer Sheva. I taught molecular structure in an intensive course as we had to cover the material of a whole term in 1 month. The lectures were in English followed by a consultation in Russian for new arrivals from the former Soviet Union. Every Israeli university invited me for a talk during that first visit. When I thought these adventures could not be topped, I received a lunch invitation to Jerusalem from Yuval Ne'eman, the minister of energy who also held the portfolio of the minister of science and technology. My hosts in Beer Sheva were quite surprised because the minister could have not been aware of every visiting professor in Israel let alone invited them for lunch. I understood that the invitation was not due to me as one of the many visitors in

science, but to our shared interest in symmetry. I hasten to add that the two of us did not represent the same weight in symmetry studies. The theoretical physicist Ne'eman was a giant who made great discoveries in the symmetries of fundamental particles. This was our first meeting in person followed by other encounters over the following years. He visited us in Budapest and we met in Stockholm when he was one of the principal lecturers in the "Symmetry 2000" meeting. In Stockholm, I recorded a long conversation with him, which appeared in the *Candid Science* series.

Yuval Ne'eman (1925–2006) was born in Tel Aviv and died there. His life could be the subject of a movie, and probably will be some day. He attended the Technion—the Israel Institute of Technology—and graduated as an engineer

© Springer Nature Switzerland AG 2020

I. Hargittai, *Mosaic of a Scientific Life*, https://doi.org/10.1007/978-3-030-34766-6_32

and physicist. He earned his doctorate from Imperial College while, as a diplomat, he was representing his country in London and a few other European capitals. Few have achieved so much as Ne'eman in physics and he could only devote half of his time to science. He held high-level functions in the Israel armed forces, especially in strategic planning. Toward the end of his career, he was much involved in domestic politics, founded a party quite on the right of the Israeli political spectrum. His role in politics was pregnant with contradiction. As it is well known, the religious parties are on the extreme right in Israel, and Ne'eman was an atheist. Their alliance was based on political expediency.

Ne'eman was no longer alive when the Higgs particle was finally observed whose existence had been predicted. Upon the appearance of experimental proofs, the theoreticians Peter Higgs and François Englert were awarded the Nobel Prize in 2013. When this happened, I remembered what Ne'eman told me about his own predictions in which he anticipated the mass of the then not yet discovered Higgs particle. He stressed this because he thought when the particle would finally be observed, his prediction might be forgotten. So I felt it my duty to remind the physicists of Ne'eman's early contribution, which was most significant. I did this not because anything might have come out of it for the late Ne'eman—there is no posthumous Nobel Prize. I wrote to Steven Weinberg as he is probably the foremost theoretician in the world in this area of physics. I felt happy when I understood that I did tell Weinberg something he had not been aware of. Of course, Ne'eman's predictions and his conclusions following from his predictions had been properly recorded in bona fide publications.

There was another Nobel Prize many years before in which Ne'eman could have had a share. In 1969, Murray Gell-Mann received it for developing the system of elementary particles and their interactions. The motivation for the Nobel Prize was carefully formulated because Gell-Mann was not the only discoverer. Independent of him, Ne'eman had also made related discoveries and nobody would have been surprised had Ne'eman been a co-recipient. In total, Gell-Mann made more discoveries than Ne'eman, but this was not or should not have been a reason for omitting Ne'eman. What may have counted more was that Gell-Mann had a special gift for coining attractive, easy to remember, and easy to pronounce names for his discoveries, whereas Ne'eman's names were precise and scientific and, accordingly, dull and difficult that would not roll off one's tongue easily.

There was a third component of the difference between these two scientists. Gell-Mann was a physicist who did physics only, all his life. Ne'eman could devote only half of his time to physics in any given period of his life. The other half was taken up by Israeli politics and defense, as well as science policy and administration. Such a multitasking approach in one's activities might make a person hurried and exhausted, but my personal experience with Ne'eman showed the opposite. He was patient and always ready for a detailed explanation. There were university aspects of his life when he founded a department at Tel Aviv University and he served the School as rector (president) for some time. It was easy to picture him as a pedagogue.

In his life, science, defense, and politics came together in unison. This found reflection in his viewing history, too. When we discussed the Jewish anti-Nazi resistance, he counted among its cases not only the Warsaw Ghetto uprising, Hanna Szenes's heroism, or other armed actions but also the participation of Jewish physicists in the Manhattan Project, the development of radar, and other war-related research activities during World War II.

Crossing Boundaries

Paul Nurse in 2003 in the author's office at the Budapest University of Technology and Economics (photograph by the author).

I first met Paul Nurse (1949–) in 2001 in Stockholm during the Nobel centenary celebrations. He was one of the new Nobel laureates in Physiology or Medicine. The award was for finding the proteins that regulate the life cycle of cells. Then, in 2003, he came for a visit to Budapest to receive an honorary doctorate from the Budapest University of Technology. Nurse has a rare comprehensive overview of modern biology and the ability to convey this information in an easy-to-perceive manner. In a lecture in 2017, he enumerated the foremost achievements of biology. Three of the five had already been around for a long time. Nurse stressed the fourth and fifth so that his audience could form a complete picture of where modern biology stood. The five are gene theory; evolution based on natural selection; that the cell is in the foundation of all life; that chemistry determines the life processes within the cell; and the central role of information in the development of biological organization.

His is a true success story considering his extraordinary elevation in British society. He received the knighthood in 1999 and from that date he is entitled to be addressed, Sir Paul. From 2003, there was an American period in his life

when he served as President of Rockefeller University in New York. He returned to England in 2010 appointed to be in charge of the new and enormous Francis Crick Institute and was elected President of the Royal Society, the pinnacle of British scientific life.

He started far below, in a poor family with a factory worker father and a cleaning lady mother—at least Nurse grew up under this notion. He was a grown-up when he learned that his older sister, whom he thought to be his older sister, was his mother and whom he thought to be his mother was his grandmother. He never found out who his biological father was. His stepfather lived long enough to know about Nurse's Nobel Prize though it was something too remote for him. It was the knighthood that the stepfather truly appreciated; it is more embedded in British life. Another Nobel laureate, John Vane, told me that he also benefited more from his knighthood in Britain than from his Nobel Prize. When he wanted to have theater tickets to a performance for which it was impossible to get tickets, he called saying that this was Sir John who wanted to get tickets and it

© Springer Nature Switzerland AG 2020   119

I. Hargittai, *Mosaic of a Scientific Life*, https://doi.org/10.1007/978-3-030-34766-6_33

worked. Had he told the theater people about his Nobel Prize, no tickets might have followed.

Nurse's siblings stopped studying at the age of 15, but his parents did not mind that he continued. It was no financial burden because Nurse always had plenty of fellowships. The parents only did not fathom what could he be still learning at the age of 30. This concern made me pause. We in academia find it most natural that we study at 30 years and older, even much older, have examinations and defend dissertations, and expose ourselves to evaluation all the time. To others, this may be puzzling.

Nurse's road to the summit of science was no easy walk. His classmates in high school came from middle-class families where they were surrounded by books, and this did not characterize Nurse's home. He was living in a permanent effort of catching up while his classmates were also making progress. His exams were not all brilliant, but it was a good lesson to get used to failures as well as success. Failures did not discourage him in his later life either. In the early stage of his research career, investigating the cell cycle was not fashionable, and he kept feeling himself as an outsider, but this was nothing new. Looking back, he saw a benefit in this, because he did not have to fight for priority in his field, and by the time it had become fashionable, he had already gained an advantage.

Based on his high school grades he was not accepted by the best schools, Oxford or Cambridge, and he enrolled at the University of Birmingham, and earned his PhD degree at the University of East Anglia in Norwich. He had a peculiar biology professor who kept questioning all the accepted "truths." Nurse thinks that all the best universities should have such an instructor. He received a good education even though his schools were not at the top of the rankings. Not only he, but his fellow graduates also proved to be competitive on the job market. Of course, Nurse's example must have exceeded any expectations.

Nurse's visit to Budapest was due to Béla Novák (1956–) who was then a professor at the Department of Agricultural Chemical Technology. From time to time he worked with Nurse in England. When I joined the University of Technology, a senior professor, Lajos Fodor, called my attention to Novák; asked me to follow his progress and assist his advancement, because Novák was very gifted. Only later did I understand that this was a sort of testament from Fodor, who was already terminally ill and soon succumbed to his illness. He felt concerned about Novák's future.

Novák grew up in a family of intellectuals and he was surrounded by a challenging environment. No effort was spared to let his talents develop. He studied biology and became a researcher of one of its most advanced branches, systems biology. This area of biology is not only about various components of any system, but also about the *interactions* between the components. This is the approach that utilizes the great advantage in considering the whole not just as a mere sum of its parts. The extra comes from considering the interactions.

I talked a great deal with Novák in the 1990s and early 2000s. I like to think that these conversations helped him in providing a sounding board for him and in his understanding that administrative positions did nothing for his advancement as a scientist. When someone is so talented, as he obviously was, he should focus on his research. He had a great feeling not only for choosing the right projects and for solving problems, but also for communicating what he had to share with colleagues in other areas of study and whose help he needed in solving complex problems. This is not trivial, because he had to involve experimental biologists and he had to illuminate them about theoretical and mathematical considerations. Then, he needed to understand what the experimentalists wanted him to know.

I had mixed feelings when the possibility of an Oxford professorship arose for Novák. I wrote one of the letters of support, but I was hoping that our University of Technology would make an offer to keep him. I often witnessed at American universities what such an outside offer can do in getting a promotion or improving one's research possibilities. I could well imagine such offers that might make staying sufficiently attractive. When we talked about what could make Novák stay, I realized that it would be impossible to fulfill. He did not exclude the possibility of staying and he did not think in terms of a higher salary or more lab space. What he wanted was a more creative research atmosphere, better conditions that result in fruitful work and interactions among colleagues. This is what he was missing and this was what he had experienced in Oxford where apparently such an atmosphere was around without making a visible effort to achieve it. Novák was 50 years old when he decided to leave and as he was departing, he told me that he would return one day. He knew about the then rigorous retirement age rules in Britain. I have known British professors who sought employment elsewhere after their mandatory retirement in Britain. Such was the example of the Nobel laureate Derek Barton. When he reached retirement age in Britain, he moved to France and when he had to retire there, he continued his active professional life in America. It may be that the British retirement laws will soften by the time Novák reaches that age, we cannot know. If he ever returns to Budapest, he will be a most useful member of our scientific community.

From Tragedy to the Summit

George A. Olah in 1995 in the author's office at the Budapest University of Technology and Economics (photograph by the author).

I was familiar with the works of George A. Olah (1927–2017) before I got into correspondence with him in 1993. He received the Nobel Prize in Chemistry in 1994. We met for the first time in spring 1995 at the Anaheim meeting of the American Chemical Society and soon afterwards in Budapest. When his school, the University of Southern California initiated the annual George A. Olah Lectures, they asked me to deliver the first one in February 1996. I talked about symmetry in chemistry in a copiously illustrated presentation in front of a large audience.

Olah's Nobel award recognized the foundation of a new area in chemistry. Its essence is that he made the rather inert carbon–carbon and carbon–hydrogen bonds capable of reacting. This led to the possibility of producing large classes of new substances that had not been possible before. There was a peculiar circumstance that accelerated the recognition of Olah's discovery. He had decided a long-standing debate between two well-known chemists. The subject of the debate was the mechanism of a specific chemical reaction. The starting ingredients and the final products of the reaction were never debated; they were reliably observed. The question was about the happenings between the start and the finish. Someone has compared it to seeing the opening and closing scenes of *Hamlet* and having to guess what happens in between. The importance of understanding what is going on during a reaction enables the researchers to influence the mechanism and moving it in directions that might lead to the production of new substances. As early as 1962, Olah already indicated that he would be able to decide the debate. In 1972, he published the basics of his new chemistry for which he received his unshared Nobel Prize in 1994. He was 67 years of age, not too old as far as recent Nobel awardees are concerned, but he had a rich career and life experience behind him.

© Springer Nature Switzerland AG 2020
I. Hargittai, *Mosaic of a Scientific Life*, https://doi.org/10.1007/978-3-030-34766-6_34

George A. Olah towering in the middle among Nobel laureates on December 10, 2001, in the Stockholm Concert Hall during the 2001 Nobel Prize centennial (photograph by Hans Mehlin; © and courtesy of The Nobel Foundation).

Olah was born in Budapest into an upper-middle-class converted and assimilated Jewish family. His father was a lawyer; they lived next to the Budapest Opera. Olah attended the Catholic high school of the Piarist Order, the *Piarista Gimnázium*. His last years in school were marred by the ever-harsher anti-Jewish laws. Hungary joined World War II on Germany's side and the worsening circumstances made his conditions not only increasingly unpleasant but also life threatening. His high school prevented persecution from entering its premises as long as it could. For example, when its Jewish students and students of Jewish origin had to be wearing the discriminatory yellow star, the school did not let them wear it within its walls. During the last months of the war, however, Olah had to find refuge elsewhere. He was saved along with a large number of other children by the Lutheran minister Gábor Sztehlo. Olah's elder brother, Peter, a graduate of the same high school, perished. Under the fiercest Nazi terror of the Hungarian Arrow Cross and the German occupiers, Olah's school remained closed from October 1944 only to reopen on March 12, 1945. Olah continued his studies and graduated at the end of the same academic year.

He continued his education at what is today the Budapest University of Technology and Economics and received his Diploma (master's degree equivalent) in Chemical Engineering in 1949. He signed up right away for doctoral studies under a professor of organic chemistry, Géza Zemplén, who encouraged independence for his disciples. Olah initiated a new field for Hungary, the chemistry of organic fluorine compounds. He prepared and defended his PhD-equivalent thesis whose materials he published mostly in Hungarian periodicals. Without losing pace, he went on with his research and prepared his dissertation for the higher doctorate, which is a prerequisite for professorial appointment. He published his new results in prestigious international periodicals. He submitted this second dissertation in 1956, but before he could have defended it, the anti-Soviet revolution broke out, was suppressed, and the Olahs left Hungary for good.

The departure of the Olah family was motivated by the hopelessness of the situation for scientific research. However, there were further considerations. It was not only Olah who suffered persecution during the wartime. His wife since 1949, Judit, was also persecuted by the Nazis. She survived only due to her reckless bravery when she stepped out of the column in which she and her sister were being dragged by the Arrow Cross to certain death. Judit escaped, her sister perished. When the Olahs were leaving Hungary, they decided never to look back to the tragic period of their lives; rather, they decided to live the life of the Americans of Hungarian origin and never let the Nuremberg anti-Semitic laws determine their lives.

They moved to Canada where family connections eased the transition. He tried but could not get employment in

When in 1965 he could return to academia, he had already moved to the USA. In 1977, he and his group transferred to Southern California; by then he could set up conditions under which he would agree to move. One of them was the establishment of an institute within the university, securing outside support, to be sure, but dedicated solely to his research. It was hydrocarbon chemistry, which Olah's discoveries developed into an exciting field. He was the first Nobel laureate of the University of Southern California. It was followed by another onc, also in chemistry, within two decades.

George A. Olah and the author in February 1996 at the University of Southern California, Los Angeles (photograph by Magdolna Hargittai).

academia. Years later after he had already earned some recognition, he received a letter from a professor of the University of Toronto. This professor begged him for forgiveness as he was the one who prevented Olah's appointment to the University of Toronto. This professor thought it too risky to have someone without a name and from a faraway place. Olah found employment in an industrial laboratory and his resolve to stay in science was tested. He had an agreement with his superiors to fulfill all his obligations to the laboratory, but having the possibility of using the laboratory equipment in his free time to continue his research. Olah organized his activities as if he was at a university department. He conducted research, invited speakers to give seminars, and published papers. It was at this time that he figured out thc solution to resolve the great debate mentioned above.

Following our initial exchange of letters in 1993, we remained in active contact with Olah throughout the rest of his life. He was like an older brother, helping me sometimes without my noticing it at the time. It was not only our shared interest in chemistry, but also in symmetry that connected us. He initiated my ever-fruitful interactions with Swedish scientists. He was supportive of my launching a magazine about the culture of chemistry, *The Chemical Intelligencer* (see the chapter Gabriela Radulescu). My journal *Structural Chemistry* organized a special issue for Olah's 90th birthday, which should have been celebrated in May 2017. The journal issue was supposed to come out in April. All participants were so enthusiastic that we managed to produce it well ahead of time. Olah could receive and browse its preprint in February, and he liked it a great deal. This was lucky, because he passed away in March.

Versatile Originality

István Orosz in 1997 (photograph by the author) and his "Complementary Kitaigorodsky" illustration for one of our books.

In his graphic art, István Orosz (1951–) often reaches out to science-related topics. He has prepared illustrations for several of our books and designed the cover for one of them. This book was about the interconnection of symmetry and scientific discoveries.[1] Each chapter carried the name of a famous scientist and Orosz prepared the chapter opening portraits. One of them, "Complementary Kitaigorodsky" is shown above. Aleksandr I. Kitaigorodsky (1914–1985) announced his research program in the early 1940s. He addressed himself to determining the most advantageous symmetry conditions in the arrangement of molecules in crystals. The far-reaching goal was to predict the crystal structures of any newly synthesized compound. There was no theory helping researchers to make such predictions and Kitaigorodsky devised an empirical route to answer this

question. He carved out a wooden model of arbitrary shape and used many such identical models for finding their most economical arrangement in three-dimensional space. He found that the best arrangement was in which the protrusion of one model fit the cavity of another—and that was a complementary way of filling the available space. Such an arrangement would not involve mirror symmetry between two models, that is, between two molecules. When there is mirror symmetry between two molecules, the cavities of the two molecules match up and so do their protrusions. Such an arrangement would not result in economical utilization of the available space. Kitaigorodsky's conclusion was that crystal structures with mirror symmetry between its molecules should be rare, whereas crystal structures in which the arrangement of molecules would be complementary (characterized by rotational symmetry) should be frequent. On the basis of painstaking examination of all possible 230 symmetry possibilities, Kitaigorodsky set up a list of

---

[1] I. Hargittai, M. Hargittai, *In Our Own Image: Personal Symmetry in Discovery* (Kluwer/Plenum, 2000; Springer, 2012).

© Springer Nature Switzerland AG 2020

I. Hargittai, *Mosaic of a Scientific Life*, https://doi.org/10.1007/978-3-030-34766-6_35

his predicted frequency occurrences starting with the highest probability and progressing toward the least probable. He did this when there were yet only very few known crystal structures. When decades later there were hundreds of thousands known crystal structures, they proved Kitaigorodsky's predictions correct. The complementary arrangement of the molecules in crystals is indeed a fundamental characteristic of their structures.

* * *

Orosz drew profound portraits for all five Martians and each is characterized by something essential. For von Kármán, he chose the human brain and space research, his Szilard's drawing refers to the book *The Voice of the Dolphins*. Szilard had a number of seminal discoveries yet it is possible that his *Dolphins* will have a longer lasting impact than his discoveries. The Wigner portrait refers to nuclear physics and von Neumann's to design and symmetry. Finally, the striking Teller portrait reminds us of Teller's involvement with the creation of the most powerful weapons that might have blown up our planet. Other characteristic drawings by Orosz show affinity with Maurice Escher's art of repeating patterns that cover the plane without gaps or overlaps. These Escher drawings have been popular in teaching crystallography similarly to the patterns of Hungarian folk motifs mentioned in the Degas chapter.

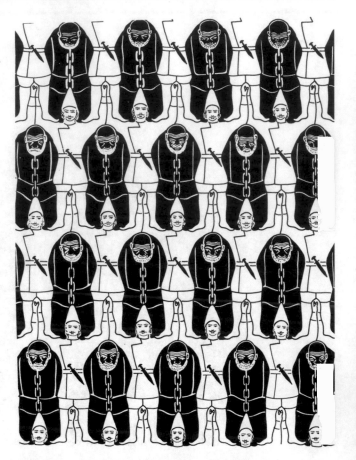

Khudu Mamdeov's "Unity" (courtesy of the late Khudu Mamedov).

Many years ago I spotted an Escher-like ad near O'Hare Airport, Chicago. I stopped my car on a busy highway and took a picture, something I would not do today, but my enthusiasm got me carried away.

The Escher-like airline advertisement with its birds flying in both directions captured the essence of its business. The images by the Azerbaijani crystallographer Khudu Mamedov (1927–1988) show kinship with Escher. Mamedov's motivation was the preservation of cultural heritage.

Mamedov studied geology in Baku, and then earned his PhD equivalent degree at the Institute of Crystallography in Moscow. In 1965, he returned to Baku and founded a structural chemistry laboratory of the Azerbaijani Academy of Sciences. He spent half a year with J. Desmond Bernal at

Birkbeck College in London. Mamedov lived at the border of two lifestyles. His parents were nomads and he observed that the nomads liked to use geometrical patterns for decoration. In contrast, townspeople liked images showing nature. Mamedov found in this another manifestation of complementary relationships. We met in 1978 at the Warsaw congress of crystallographers. He displayed his periodic drawings on a small table and generated a great deal of interest among the attendees of the meeting. Many wanted to buy his art, all of which were hand-drawn unique pieces, but he was not interested in selling. We talked quite a bit, and he gave me a few of his drawings at the close of the congress. In 1982 I visited him in Baku. In his laboratory, his home, and his home town, where he took me, I experienced a mixture of modern science and traditional lifestyle.

His drawing "Unity" represents old warriors and young fighters. The faces of the young are almost uniform; the old veterans show more individuality. Apart from the facial expression, all young men and all old men appear to be the same. A few years after Mamedov's death, his disciples published a luxurious volume with his drawings that were not facsimile copies; rather, they were computer-generated patterns. In this version of "Unity," not only the young had uniform facial expressions, but the old as well.

* * *

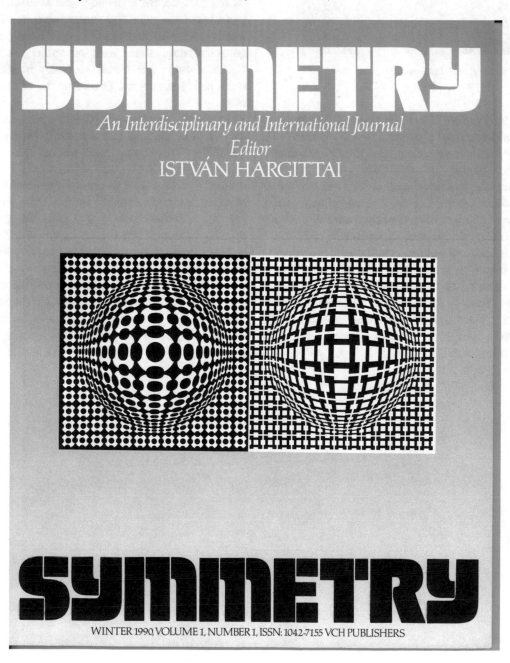

Anti-symmetric Vasarely pattern on the cover of the *Symmetry* journal.

Ernő Lendvai (photograph by the author).

Victor Vasarely (1906–1997) was one of the pioneers of op-art. His works are rich in periodicity and symmetry. He was born in the southwestern Hungarian town Pécs and died in Paris. He has permanent exhibitions both in Pécs and Budapest where his traditional and avant-garde pre-op art paintings are also of great interest. I started corresponding with him when in the first edition of *Symmetry through the Eyes of a Chemist* we wanted to reproduce one of the drawings of his series about the creation of the world. He was most accommodating. Alas, this image did not figure in subsequent editions of the book, because his heirs were less accommodating. Back in 1990, when I launched my short-lived *Symmetry* periodical, I published a short paper explaining the symmetries and anti-symmetries in his works. We corresponded in Hungarian and he showed

interest in a more detailed analysis of the symmetries and anti-symmetries in his pictures. He sent me a lot of reproductions and a book. I enjoyed looking at his pictures, but did not feel prepared to analyze them in greater detail.

* * *

Symmetry survived only its first issue. As I was correcting the proofs of the second issue, I learned about the sale of the publisher and the termination of the journal by the new owner. One of the losses due to this abrupt end was the second part of Ernő Lendvai's (1925–1993) comprehensive paper about symmetries in music. He was an internationally renowned musicologist especially for his studies of Béla Bartók's music. His knowledge and enthusiasm were contagious and I also did some measurements of a number of Bartók pieces to find out for myself the presence of the golden section. There were numerous Bartók recitals in Budapest during the Bartók centenary in 1981 and Magdi and I indulged in them. Otherwise, he is not performed so often.

Bartók was much discussed in the media during the centenary and I contributed to a discussion. An article raised the possibility of performing music without identifying either the composer or the performers. In response, I wrote: "Bartók stressed that we should direct our interest at the actual work of art rather than focus on the name of its creator. He cited the joy that is derived looking at a cathedral or a painting or listening to a poem without having any knowledge about the architect, artist or author. He was wondering whether it might be advantageous to perform musical works without any mention of the composer's names. This was but another expression of his legendary modesty and dedication, some of the many qualities of his human greatness. . . . knowing about Bartók may add to the pleasures of listening to his music. But, surely, it is his music that comes first."[2]

[2] I. Hargittai, "First the Music." *International Herald Tribune*, August 6, 1981.

Walking into Friendship

Left: Guy Ourisson and the author in 2000 at the French Academy of Sciences (by unknown photographer; courtesy of the late Guy Ourisson). Right: Jean-Marie Lehn in 1999 in Budapest (photograph by the author).

Guy Ourisson (1926–2006) researched chemical compounds occurring in nature and discovered many previously unknown substances. However, among his many discoveries, he was the most proud of having "discovered" Jean-Marie Lehn (1939–). In the early 1990s, I attended a meeting on organic chemistry in Bürgenstock, Switzerland. I took long walks and Ourisson also enjoyed long walks. This is how we met so we had a joint walk for a couple of hours during which a rain shower caught us but we did not pay attention. We talked, exchanged thoughts and ideas, got an insight into each other's lives, and for these couple of hours we became the best of friends. The conversation was apparently so comprehensive that very little else remained afterward, except the glow of this most pleasant exchange. For me, this was what I remembered from this meeting in Bürgenstock.

Ourisson was born in Boulogne-Billancourt, France, graduated from the famous École Normale Supérieure in Paris, and earned his PhD degree in 1952 from Harvard University. Upon his return to France, he acquired his higher doctorate in 1954 and had a usual academic career from 1955. He capped it in an unusual way as he was the founding Rector (President) of the Louis Pasteur University in Strasbourg. He had high positions in French academia and science politics.

When Ourisson was elected president of the French Academy of Sciences, he invited Magdi and me for a visit. He was most attentive in all the details. He reserved a room for us at the "Hotel des 3 Collèges," a hotel with past Hungarian connections. Its façade displays a memorial plaque of Mikós Radnóti, my favorite poet, who was a slave laborer

© Springer Nature Switzerland AG 2020

129

I. Hargittai, *Mosaic of a Scientific Life*, https://doi.org/10.1007/978-3-030-34766-6_36

Miklós Radnóti 's memorial plaque on the façade of 16 rue Cujas in Paris (photograph by the author).

in 1944 and shot to death by officers of the Hungarian Army. The plaque quoted the concluding lines of his poem "Hispania, Hispania." In English translation[1]:

> Freedom, your fate is in peoples' throats!
>> it was for you, that song this afternoon;
>> with weighted words they sang your war,
>> with their drenched faces, the Paris poor.

As a guest of the French Academy, I received various honors and was introduced at the meeting of the Academy. While the introduction was being read about my activities, I was informed that I should respond with a brief speech. I am usually fond of unexpected situations. For example, I prefer direct broadcast of radio interviews to ones that can be restarted several times for later editing. Yet this was a little too unexpected. Fortunately, I noticed among the French academicians a few to whose research I could relate my own interests and it all came out well. We spent a great deal of time with Ourisson, but the Bürgenstock atmosphere of "discovering" each other never returned.

Ourisson had an interesting background. His father was a Jew from Łódź, Poland, and his mother was the daughter of a French nobleman and a Russian princess, a descendent of the Ermolovs. One of her ancestors conquered the Caucasian region for the Czar and fought in the mountains of Chechnya. Ourisson grew up in a chemical plant and played with chemicals, something nobody today would allow children to do. He was 24 years when he embarked on his doctoral studies at Harvard University. Due to a misunderstanding he thought that he would have 2 years for completing his work. He succeeded, and it has stayed as a record at Harvard. Looking back, he regretted the haste and that he did not make better use of his stay there.

In his research, he used enormous quantities of the starting material to reach minute amounts of the most valuable ingredients. His projects originated from the former French colonies and often local wars and world politics interfered with his work. For example, his rich Cambodian sources dried up due to the Vietnam War. In addition to his fundamental research, he did a great deal of contract work for companies and he used his earnings for funding prizes and stipends.

Being president of the French Academy of Sciences, Ourisson was much occupied with the relatively poor standing of French science on the world scale; for example, by the scarcity of Nobel Prizes for French scientists. He thought it was wrong to centralize French science, to isolate it from the rest of the world, and to enforce the French language in scientific communication. In his opinion, the difficult route

---

[1] *Miklós Radnóti: The Complete Poetry*. Edited & Translated by Emery George (Ann Arbor, Michigan: Ardis, 1980), p. 211.

for gifted young scientists to gain independence was also a retarding factor. He was proud that he recognized early Jean-Marie Lehn's talent and helped him to become independent. Lehn's Nobel Prize and his personality have done a great service to French and universal science and to France in general. In Lehn's case the Nobel Prize has multiplied his possibilities of helping science, including chemistry, which badly needs improvement of its image, often associated with pollution.

Ahead of His Time

Michael Polanyi in 1931 in Berlin
(courtesy of John C. Polanyi).

Michael Polanyi (1891–1976), Hungarian-British scientist, was a physician turned physical chemist turned philosopher. He made forward-looking discoveries in understanding the phenomenon of gas adsorption, pioneered a new direction in X-ray crystallography, and his best-known chemical discoveries concerned the mechanism of chemical reactions. His most influential work in philosophy was his 1958 epistemological book, *Personal Knowledge*. He created a unique contribution to twentieth-century science and culture. The Jewish Polanyi converted in his youth, but never denied his Jewish roots and stressed it especially in the critical early 1940s.

I met Polanyi when I was a visiting research associate at the physics department of the University of Texas at Austin in 1969. He came for a visit and the chairman of the department, Harold P. Hanson, took Polanyi and me to the plushest Austin club for lunch. At the time I was familiar only with Polanyi's discoveries in physical chemistry and knew nothing about his philosophy. Our conversation covered broad grounds; for example, foreign words in the Hungarian language; philosophy; history; and the information explosion, which then was not yet so striking as it is today. Polanyi issued stern warnings as to its drawbacks in addition to its obvious benefits. Both the scope and the atmosphere of our conversation left a deep impression on me. He spoke quietly, unhurriedly, and factually. Only someone truly knowledgeable can be so convincing while staying unpretentious.

On October 5, 2016, the Fritz Haber Institute of the Max-Planck-Gesellschaft in Berlin organized a day of lectures in remembrance of its former associate, Michael Polanyi. It was in celebration of the 125th anniversary of his birth. There were three lectures and a few short contributions. The American Nobel laureate Dudley Herschbach spoke about some of Polanyi's favorite issues, tacit knowledge and keen insight; one of Polanyi's biographers, Mary Jo Nye, about personal knowledge and social practice; and I spoke about Polanyi as an "Honorary Martian of Science." There was a letter from Polanyi's son, John C. Polanyi, and a comment by the German Nobel laureate Gerhard Ertl among the short contributions.

There were many similarities in the family background of Polanyi and the "Martians," and many points of intersections in their lives. Polanyi attended the Mintagimnázium (Model HS) in Budapest, like Theodore von Kármán and Edward Teller. Polanyi's sister, Laura, studied in the Lutheran Gimnázium, like Eugene P. Wigner and John von Neumann (though Laura was a "private" student as the school was for boys only). In the early 1920s, Polanyi mentored Wigner in

Berlin for Wigner's doctoral studies. According to the science historian Abraham Pais, Polanyi had a decisive impact on Wigner's thinking not only in physics, but also in philosophy and politics.

For his achievements in science, Polanyi could have been a member of the "Martians," but he did not participate in the war efforts like they did. He intended to do defense-related work and offered his services to the British authorities. They declined because they did not recognize the relevance of his research to the war efforts. He was involved only indirectly. When John A. Wheeler was working out the detailed theory of nuclear fission for the atomic bomb (in cooperation with Niels Bohr), he was relying on Wigner and Polanyi's prior theoretical studies. According to Wigner, Polanyi "moved to Manchester, England, in 1933, when Hitler came to power, *for a reason very similar* to that which had originally prompted him to leave Hungary."[1] (my emphasis) According to some, Polanyi and many others were not forced to leave Nicholas Horthy's anti-Semitic and autocratic Hungary. Wigner saw this differently; hence his comparison of Hungary of the 1920s to Germany of 1933.

Polanyi graduated from high school in 1908 and received his MD diploma in 1913. He served as a military physician in the Austro-Hungarian Army in World War I. He began his scientific research during his medical studies. His discoveries on the adsorption of gases served as his PhD dissertation. His supervisor sent the dissertation to Albert Einstein, who liked Polanyi's work.

Polanyi had an important function in higher education in the ministry of education for both the 1918 civilian and the 1919 communist administrations. His role was professional and not political. He condemned the so-called white terror following the suppression of the communist dictatorship. He recognized that Horthy's regime had no place for young, ambitious talents, especially not if they were Jewish. He left Hungary for Germany, which, at the time of the Weimar Republic, was a democratic country. He retained some interactions with a few colleagues in Hungary and did external work for the Tungsram Company. However, he had little interaction with official Hungarian academia and he was never elected to membership in the Hungarian Academy of Sciences. In 1944, he was elected Fellow of the Royal Society (FRS, London).

In the late 1920s, Polanyi was accused in the Hungarian press of denying of being Hungarian. Polanyi explained painstakingly the reasons for his emigration. He mentioned the anti-Semitic law of 1920, which limited the number of

Jews accepted by Hungarian universities, known also as the law of *numerus clausus* (closed number). He stressed that lacking proper perspective encouraged not only Jewish youths to depart, but others as well.

In Germany, Polanyi became a successful scientist. He was watching the rise of National Socialism, but for a long time could not have imagined that the "brown madness" would overtake a country of such great culture as Germany. When he realized that he was mistaken, he left for England and received a professorial appointment in Manchester. There, he created an internationally renowned laboratory of physical chemistry.

John C. Polanyi in 1995 at Toronto University (photograph by the author).

He had excellent associates and disciples. One of them, the American Melvin Calvin, would later receive the Nobel Prize for his discoveries in photosynthesis. Calvin started his research in this area while working for Polanyi in Manchester. Polanyi's Nobel laureate son, John C. Polanyi, was not Polanyi's pupil, but he learned a lot of chemistry from his father's pupils and much else from his father. According to John, his own Nobel recognition brought, in some way, Michael Polanyi to Stockholm. Alas, at the time

---

[1] *The Collected Works of Eugene P. Wigner, Volume VII: Historical and Biographical Reflections and Syntheses.* Annotated and Edited by Jagdish Mehra (Berlin, Heidelberg, New York: Springer-Verlag, 2001), p. 154.

of John's Nobel Prize, Polanyi had already been dead for 10 years.

The successful physical chemist Michael Polanyi made another big career change in 1948. He moved to philosophy, still at Manchester University. He became influential, especially in epistemology—the theory and practice of acquiring knowledge. He was interested in studying the behavior of scientific researchers in various political systems and condemned those that succumbed to the false Soviet propaganda. He noted that the values and benefits of democracy may be underestimated easily by those who were living in it.

When I examined the archives of the former Hungarian secret police, I found documents from the late 1950s and early 1960s witnessing an enhanced interest of the secret police in the "loyal Hungarian émigrés in England." There are reports about successful Hungarian doctors, lawyers, writers, artists, economists, engineers, and scientists who immigrated to Great Britain, most fleeing the anti-Semitic Horthy regime. Outstanding contributors to world culture figured among them, such as the future Nobel laureate inventor of holography Dennis Gabor, the economists Nicholas Kaldor and Thomas Balogh, and Michael Polanyi.

Polanyi made outstanding scientific discoveries, but when truly new findings emerge, their recognition is often hindered by their very novelty. This was characteristic of how Polanyi's discoveries were met. His situation was yet more difficult as he arrived from other areas than the ones in which he made his discoveries. He had to establish himself in chemistry coming from a physician's background and then in philosophy, coming from physical chemistry. John Polanyi has called my attention to this paradox that Polanyi's original training was in medicine where he did not make discoveries. His discoveries were all born in fields where he arrived as a guest and where he began his activities as an outsider. Yet he made in each of his chosen areas milestone discoveries.

A scientific discoverer is lonely by the nature of making a discovery, because he or she possesses a knowledge that nobody had possessed before. This loneliness characterized Polanyi even to a greater extent than most. Yet he did not become bitter and did not give up making new and newer discoveries. On more than one occasion he was ahead of his time; his discoveries proved "premature." There are discoveries that are so timely that the scientific community is ready to accept them as soon as they appear. For other discoveries, a longer period is necessary before they can be built into the texture of contemporary science. Yet other discoveries are so much ahead of their time that the scientific

community does not digest them and they may even be forgotten for a while, only to be rediscovered at a later stage. Polanyi's discoveries on adsorption in the 1910s were such discoveries and it was even more unfortunate that this happened at the beginning of his scientific career.

He had the talent of making discoveries soon after he had joined a new area of research. This is what happened when he started doing X-ray crystallography in the 1920s in Berlin. He recognized the significance of the structure of fibrous materials. His were pioneering observations and suggestions for further work, having importance for the elucidation of biological macromolecules. Yet this all happened at the time when it was still being debated whether or not those biological macromolecules existed in the first place.

At the time of Polanyi's involvement with X-ray crystallography, the target of studies was the ideal crystalline materials. Hardly anybody recognized the importance of fibrous materials that would bring great breakthroughs in science during the second half of the twentieth century. Polanyi's choices of research projects were never influenced by "fashion," and he focused his X-rays studies on fibrous materials.

The internationally renowned mathematician George Pólya (1887–1985) must have anticipated Polanyi's future difficulties when he warned him that if he would go his way alone, he would need a strong voice to make himself heard. Polanyi was a superb mentor and he paved the way most efficiently to great careers for several scientists. He himself though had no mentor in his fields of inquiry. Fourteen of his first fifteen papers he wrote and communicated without a co-author. Robert Frost's words fit Polanyi's self-chosen fate,[2]

> Two roads diverged in a wood, and I—
>   I took the one less traveled by,
>   And that has made all the difference.

Polanyi's choices of his topics of investigation were governed by his interest and nothing else. His greatest success in science was in uncovering the mechanism of chemical reactions, in the understanding of what stabilizes molecules and what makes them react, and what happens to them during their reactions. His descriptions of his findings were not only correct they were also attractive aesthetically, even if such an adjective sounds unusual in connection with a scientific theory. His career, his demeanor and humanity exemplify the scientist who is willing to question dogmas and demonstrate the usefulness of being engaged in more than one culture.

---

[2] Robert Frost, "The Road Not Taken," the last three lines (https://www.poetryfoundation.org/poems/44272/the-road-not-taken); downloaded May 19, 2019.

A Short-Lived Magazine

Covers of *The Mathematical Intelligencer* and *The Chemical Intelligencer*. The top two represent geometrical constructions related to the buckminsterfullerene molecule (C_{60}) in which regular hexagons with embedded pentagons form sphere-like shapes (the photographs are by the author). Bottom left: the double helix sculpture of Bror Marklund at Uppsala University (photograph by the author). Bottom right: Quasicrystals of aluminum/manganese alloys, prepared by Ágnes Csanády; electron micrographs by Hans-Ude Nissen.

Gabriela Radulescu is an outstanding American media expert who has founded and managed journals and with whom I was in contact during the publication life of the magazine, *The Chemical Intelligencer*. The archaic word intelligencer means reporter and the magazine was about the culture of chemistry. I initiated the magazine and edited it during its existence, 1995–2000. At that time, it was not too common that a magazine published in New York would be edited in Budapest. It ceased to exist due to mergers and acquisitions of publishing companies.

When I suggested creating such a magazine, I modeled it after *The Mathematical Intelligencer* to which I contributed frequently. The Publisher responded favorably almost as if I had come up with an original idea although they had been publishing the mathematical magazine for a long time. My contact was Radulescu and we interacted for years, but met in person only once when the magazine no longer existed and Radulescu had left the company.

The organization of the magazine began with my sending out about 30 letters of inquiry to well-known chemists soliciting comments on the idea of establishing such a periodical. One of them was George A. Olah. This was in 1993 and this was a fortunate start of our interactions, 1 year *before* his Nobel Prize. Another was Linus Pauling who indirectly gave me the idea of interviewing scientists for the magazine and the one with him appeared in the inaugural issue. When I collected a good number of supportive statements, Radulescu prepared a plan for the magazine in which she spelled out the support that the Publisher offered. I responded with enthusiasm. In her immediate response, she asked me to prepare another response to her initial offer in which I would express my reaction in a more subdued manner. My letter might give the wrong impression that she might have given me a too generous offer.

This episode reminded me of an earlier experience and indicated that I had not learned its lesson. In 1984, following my first year of visiting professorship at the University of Connecticut at Storrs, Dean Julius Elias (1926–2008) asked me to stay for another year. I was happy. I worked hard during my first year; I had never taught in English before, and on top of this, I had to teach physics, which I had never done before either. Originally my invitation was for doing research only, but by the time we were ready to leave from Budapest, with our two children, it turned out that the university needed me to teach a subject. Only at the last moment did I learn that it would be physics. It was one of the hardest challenges I had ever faced. The dean appeared to be happy with my performance and offered a substantial raise. I did not hide my satisfaction and thanked him. His reaction surprised me. He told me that I took the wrong approach for such a discussion. I should rather say that I am happy to be asked for a second year, but the raise was insufficient considering the task and would like a higher remuneration. As he was looking at me with curious anticipation, I told him that I was happy to be asked for a second year, but the raise was insufficient considering the task and would like a higher remuneration. He said that they were happy with my performance and the raise he offered reflected this, but their offer was the maximum they could do. Again, I sensed his anticipation of my reaction. I repeated what I thought he had taught me to say, and he again assured me that they could not offer me more. We repeated this exchange and I started feeling uncomfortable. The last thing I wanted was risking my second-year appointment. Finally, he told me that I learned the lesson, and doubled the raise.

Returning to *The Chemical Intelligencer*, it soon became quite popular. One of our subscribers was the Nobel Committee of Chemistry and world-renowned scientists were members in our Editorial Board and among our authors. The circle of subscribers was small; it was growing, but slowly. The Publisher initiated a study whose prognosis was optimistic and *Nature* published an encouraging review of the first-year issues. Then, in the sixth year of the magazine, its old, prestigious publisher was sold and the magazine was terminated. The cover of the last issue showed the covers of the 23 previous issues. In 2015 my son, Balazs, and I published a volume of selected articles (for the cover of the volume, see the Ferenc Lantos chapter). The book was well received, which could be taken as a sign of nostalgia toward the magazine. For me the termination of the magazine may have been a positive development in disguise; it was taking an increasing amount of my time as I, alone, was doing the editorial work. A number of projects have originated from *The Chemical Intelligencer*. It was the beginning of quite a few enriching personal interactions.

I met with Radulescu when she was already a freelancer. She had initiated a new magazine about women's health, which had an attractive program, and we subscribed to it right away. Alas it was a one-person enterprise and soon it closed. As a parting act, we had dinner together in Chicago. Radulescu was full of ideas about science publishing and more generally about communications, and she had the ability to articulate her plans in a subdued, but convincing manner. At that time, I had Japanese friends who appeared to have financial backing for launching a new English language scientific magazine in Japan. They were ambitious and wanted to create an equivalent of a *Science* or a *Nature*. They were prepared to run the show even for 100 years before expecting it to turn profit. It was prestige they were looking for. I suggested to them to team up with Radulescu who found the concept to be of interest and was willing to be tested by such a project. Unfortunately, a recession killed the Japanese plans.

Worthy of His Fate

Andrei Sakharov's statue (by Levon K. Lazarev, 2003) on Sakharov Square in St. Petersburg with university buildings in the background (photograph by the author).

The theoretical physicist and human rights activist Andrei D. Sakharov (1921–1989) is one of a few in this book about whom I write without having ever met him in person. I have known a number of his family members, one of his daughters, two of his sons-in-law, and one of his granddaughters. I have read everything I could find by Sakharov and about Sakharov. Magdi and I have visited the Sakharov Archives and the Sakharov Center in Moscow. These are independent institutions and we studied their collections. I have been profoundly impressed not only by the greatness of this man, but even more by the changes over the years, how his interest in and dedication to human rights kept progressing. He was one of the 12 principal characters in my book *Buried Glory* about Soviet scientists.

© Springer Nature Switzerland AG 2020

I. Hargittai, *Mosaic of a Scientific Life*, https://doi.org/10.1007/978-3-030-34766-6_39

The cover of *Buried Glory*.

I was a visiting scientist at Oslo University in 1975 when the Norwegian Nobel Prize Committee for Peace awarded the Nobel Peace Prize to Sakharov. I was much moved by the spontaneous torchlight demonstration in downtown Oslo in the evening of the prize awarding ceremony. This was not a demonstration organized by the authorities that I was used to in Hungary at the time. Sakharov was not in Oslo; the Soviets did not let him out of the country. On his behalf, his wife, Elena Bonner, received the award.

I read the following Sakharov statement in the Moscow Sakharov Archives (my translation):

> . . . my fate was larger than what would have followed from my personality. I was merely trying to be worthy of my fate . . .

I found this statement most characteristic, because knowing about his path the impression is that he indeed merely followed what fate had charted for him. Sakharov inherited his interest in physics from his father and his religion from his mother. He abandoned religion in his youth, but preserved his respect for the believers for his entire life. He studied physics at Lomonosov University. The School was evacuated to Turkmenistan during the war, but the level of instructions remained excellent. Sakharov kept his distance from ideology and from political organizations, not by design, but because it just happened. He was rather an introvert, but very active in suggesting innovations already in his first workplace, a plant for manufacturing ammunition.

He joined the Physical Institute of the Soviet Academy of Sciences after the war and became a member of Igor Tamm's theoretical group. As one of Tamm's associates, he moved to the secret nuclear laboratory, Arzamas-16, and stayed on there even after Tamm had returned to Moscow. He was a leading theoretician of the nuclear program and he proposed two of the three principal ideas for developing the Soviet hydrogen bomb. The third principal idea was Vitaly

Ginzburg's, who, however, stayed in Moscow and was soon removed from the program on account of his exiled second wife. Ginzburg was considered unreliable.[1]

Sakharov became a member of the Academy of Sciences early and was showered with the highest awards for his participation in the nuclear program. His first clash with the authorities came about in 1955. There was a celebration of a successful test and in his toast he expressed his desire that their bombs explode always over proving grounds and never over cities. The representative of the Soviet government understood that Sakharov wandered into the political arena and he warned the physicist to stay with science. Then, in 1961, at another celebration Sakharov argued that it was superfluous to build bombs of ever-increasing power, and pointed to the health hazards of the experimental explosions. On this occasion, Nikita Khrushchev himself reprimanded Sakharov in a most humiliating manner. Sakharov was studying the possible genetic harm of the testing and it was gradually dawning on him that the Soviet authorities ignored the risks to which the population might be exposed. Paradoxically, however, he was also looking for ways to cause the utmost damage to the enemy by the deployment of the horrible weapons he had helped to develop. I mention this, because in light of later development, it might appear unbelievable that Sakharov would advance such schemes, but according to his own memoirs, he did.

He returned to Moscow in the 1960s, stopped working on nuclear weapons and moved into fundamental research in which he also excelled. In parallel, he was gradually becoming involved in human rights activities. He collected signatures for petitions protesting the exoneration of Stalin, attended protests in front of courthouses where human rights activists were being tried, and so on. The Police arrested other protesters, but they did not touch Sakharov, the thrice Hero of Socialist Labor and the "father of the Soviet hydrogen bomb," the cornerstone of the Soviet superpower status. His frail physique lent tremendous power to his appearance in these dangerous situations. By 1980, the authorities were fed up with his ever-increasing influence and exiled him to the closed city of Gorky (before, and now, again, Nizhny Novgorod).

The KGB constantly harassed him in his exile. When he reached a certain stage in writing his memoirs, his manuscript was stolen from him, and he had to start it all over again. When he went on hunger strike in protest against various machinations of the authorities, he was force fed and his life was threatened by the treatment. Whatever the authorities tried doing to break him, his resolve kept strengthening. His Moscow colleagues had permission to visit him annually and informed him about new developments in physics. The Physical Institute still had his name on its list of associates and one of the offices was still assigned to him, thereby, if only symbolically, counting him as one of them. Even under these impossible conditions, Sakharov kept to the rules he had accepted when he was a member of the nuclear program. On an occasion when he and his visitors were discussing what might have been classified information, Sakharov stopped the discussion. He said that although they all had the necessary security clearance, the KGB officers listening clandestinely to their conversation might not.

Mikhail Gorbachev had already been in power for 18(!) months when he finally let Sakharov back to Moscow. Sakharov returned to his research, but during the last 3 years of his life he was increasingly involved with politics. He acquired worldwide popularity and when he died the world mourned his loss.

---

[1] Igor Tamm was a co-recipient of the Nobel Prize in Physics in 1958 and Vitaly Ginzburg in 2003. Ginzburg's wife was exiled from Moscow under false accusations about her alleged plans to assassinate Stalin.

Art of Interviewing

Géza Simonffy in 1982 from the archives of Radio Budapest (http://www.fortepan.hu/_photo/download/fortepan_56188.jpg).

Géza Simonffy (1923–2005) was, for many years, in charge of the section for popular science at Radio Budapest. Our paths crossed in September 1965 when he asked me to record a conversation with the Soviet Nobel laureate, Nikolai Semenov, who was coming for a brief visit to Budapest. I accepted the invitation without hesitation. It did not even occur to me that I had never taken any interview before. I spoke Russian, Semenov's name was familiar to me, I was a chemistry researcher, and found nothing extraordinary that I was asked. The Radio assigned a technician with a suitcase-size recorder to the project, and we visited Semenov in his hotel room.

In hindsight, I was hopelessly unprepared for this interview, but Semenov was a seasoned interviewee. An interesting conversation was the result; it was broadcast several times, and its transcripts appeared in a volume of the best programs. I published an article about Semenov's discovery of the mechanism of the branched chain reactions, which he discovered around 1930, and for which he received a share of the 1956 Nobel Prize in Chemistry.

For me, the most interesting part of the conversation was his narration of the conditions for science in the Soviet Union of the 1920s. New research institutes mushroomed; members of the society who could have not dreamed of becoming scientists could attend university. They started research parallel with their studies. There was euphoria to create a new Soviet science and a new Soviet intellectual class. Of course, Semenov idealized everything and did not say a word about the destruction of the "old" intellectual class. In this case, destruction should be understood verbatim. The lucky ones were exiled to the West, many others lost their jobs and means of doing science, and quite a few were murdered. This was not in the interview and all this I would only learn about years later.[1] Semenov's enthusiasm was genuine. He had a brilliant career in those 1920s and later, and stayed an unquestioning pillar of the Soviet power to the end of his life. He was also a gifted scientist who established a large network of institutes and co-workers whom he encouraged to find their independent projects.

Simonffy was happy with the outcome of the Semenov interview. I did a few more during the subsequent months though none as successful as the Semenov conversation. Then I stopped doing interviews for the next 30 years, because I focused on my research. For my magazine, *The*

---

[1] See, e.g., I. Hargittai, M. Hargittai, *Science in Moscow: Memorials of a Research Empire* (Singapore: World Scientific, 2019).

Chemical Intelligencer (see the Gabriela Radulescu chapter), I started interviewing again. Even when the magazine folded, I continued doing so for a while. In the early 2000s, my wife and I started compiling the volumes of the *Candid Science* series of interviews, and our son, Balazs, joined in. It was then that I remembered the Semenov interview, and purchased a copy of its tape from the Archives of the Radio. I

the Foreign Ministry, and most of all, the KGB. Nonetheless, Semenov's award was justified. It was a sad irony of history that the first Soviet science Nobel Prize coincided with the Hungarian revolution and its brutal crushing by the Soviets in fall 1956. The Swedes organized the award ceremony in a subdued atmosphere, thereby expressing their solidarity with the Hungarian cause.

The author at Nikolai N. Semenov's grave in 2011 at the Novodeviche Cemetery in Moscow (courtesy of Larissa Zasourskaya).

translated the conversation into English and included it in the first *Candid Science* volume. Years later, a Moscow publisher brought out the first two volumes of *Candid Science* in Russian translation. The translator sent me the manuscript for checking. It was interesting to see the Semenov interview translated from the English translation back into Russian. It was fully consistent with the original Russian.

Semenov's Nobel Prize was the first for a Soviet scientist. The Swedes had long tried to involve the Soviets in, as they put it, "the Nobel movement." In their effort, they overstepped some of the rules they usually enforce. It is stipulated that only individuals and not organizations or institutions, can make nominations for the science prizes. In contrast, on the Soviet side, no nomination could have gone through without the involvement of the Science Academy,

Semenov's grave is in the prestigious Novodeviche Cemetery in Moscow. The tombstone is a larger than life-size statue erected by his third wife. Semenov's first wife was 17 years his senior. When they met, she was married with children, a vivacious lady who was the center of any company. Soon after she married Semenov, she became ill and died. Before her death, she designated Semenov's second wife in her own niece. In the second marriage, there were two children, and it was a harmonious union. However, during the celebration of Semenov's 75th birthday, he stunned his wife and the guests by declaring that he was filing for divorce, and he left the party. His third wife was decades his junior, an associate of the institute under his directorship. According to the members of the Semenov family, the tombstone statue is not lifelike; it presents

Semenov as detached and aristocratic, whereas he was friendly and plebeian.

When I resumed interviewing in the 1990s, I gradually worked out my technique. I learned a great deal from the preparations and even more from most of the actual conversations. I was preparing consciously, but only up to a

video interview program, and collected some 80 interviews. They never transcribed the conversations and never published anything from their material. Balazs and I selected a few and included them in the fifth volume of the *Candid Science* series.

On rare occasions, I interviewed without proper prepara-

Glenn T. Seaborg and the author at the Springer-Verlag booth during the spring 1995 Anaheim meeting of the American Chemical Society (by unknown photographer).

point. My experience was that a "limited preparedness" may be optimal for the success of the conversation. Being unprepared was unthinkable. But knowing everything possible about the interviewee in advance dampened my curiosity. It was also instrumental to stress that I was not a journalist; rather, a fellow scientist. This facilitated the interviewee to open up. I promised that nothing of the conversation would be printed before the interviewee would have the possibility of checking, correcting, replacing, deleting, and augmenting the text of the conversation. I transcribed all my interviews, however time- and labor-consuming this was and sent the slightly edited transcripts to the interviewee.

Eventually, these interviews became known and welcomed. When *The Chemical Intelligencer* was terminated at the end of 2000, my interviews appeared for a few more years in the magazine *Chemical Heritage*. We received the highest recognition for our interviews when after Clarence Larson's death, his widow, Jane Larson, donated their rich collection of video interviews with scientists and technologists to us. Clarence was a chemistry professor and science administrator, and a former commissioner of the US Atomic Energy Commission. When he retired, he and his wife initiated a

tion. Thirty years after such an interview with Nikolai Semenov, it happened again in 1995. During the spring meeting of the American Chemical Society, as I was walking in Anaheim, I bumped into Glenn T. Seaborg (1912–1999). On the spur of the moment, I introduced myself and asked him for an interview. He graciously agreed, but the only available time for him was right then. Fortunately, I had a miniature recorder with me even if no camera. We found a quiet corner and recorded a conversation. Again, like Semenov had, Seaborg saved the situation having been a seasoned interviewee. The conversation appeared in 1998 in *The Chemical Intelligencer*. Still, I always felt that the Anaheim conversation was incomplete, so still in 1998 I asked him for some addition. While in 1995 the naming of the element 106 was still under debate, in 1997, it was finally resolved and it became Seaborgium, Sg. The augmented interview appeared in 2003.[2]

---

[2] Istvan Hargittai, *Candid Science III: More Conversations with Famous Chemists* (edited by Magdolna Hargittai, London: Imperial College Press, 2003). Chap. 1: Glenn T. Seaborg, pp. 2–17.

Concluding volume of the *Candid Science* six-volume set of interviews with famous scientists.

Altogether six volumes of *Candid Science* appeared at Imperial College Press in London. Each volume contained at least 36 conversations with at least half of them with Nobel laureates. The number of Nobel laureates has increased since, because so far, a dozen interviewees have received this distinction afterward. When Magdi noticed that too few women appeared in our volumes, she reached out and focused on famous women scientists. The result was her book about women scientists.[3] Recently, we chose 111 of the *Candid Science* conversations and compiled extracts from them in a separate volume for a broader readership, under the title *Great Minds*.[4]

[3] Magdolna Hargittai, *Women Scientists: Reflections, Challenges, and Breaking Boundaries* (New York: Oxford University Press, 2015).

[4] Balazs Hargittai, Magdolna Hargittai, and István Hargittai, *Great Minds: Reflections of 111 Top Scientists* (New York: Oxford University Press, 2014).

He Braved History

Albert Szent-Györgyi in 1947 in Cambridge, England, with the future Nobel laureate crystallographer Dorothy Hodgkin (courtesy of William B. Jensen and the Oesper Collection of the University of Cincinnati).

Albert Szent-Györgyi (1893–1986) is yet another of the personalities in this book whom I never met in person. I knew one of his close associates, Ferenc Guba (1919–2000), and his cousin, Andrew Szent-Györgyi (1924–2015), who was close to him in America. Guba and I worked in the same laboratory for a few years in Budapest. Then, Guba moved to Szeged when he was appointed to Szent-Györgyi's old chair at Szeged University.

Szent-Györgyi was awarded the Nobel Prize in Physiology or Medicine in 1937 for his discoveries related to biological combustion and Vitamin C. He conducted a considerable portion of his prize-winning research in Hungary and has been the only Hungarian Nobel laureate in science who traveled to Stockholm from Budapest to receive the award. He was not yet a full member of the Hungarian Academy of Sciences at the time of his Nobel Prize. He was first proposed for corresponding membership in 1931, but was not elected although he was by then an internationally renowned scientist. He became a corresponding member in 1935 and full member in 1938, already as a Nobel laureate.

Szent-Györgyi was 41 years old at the time of his corresponding membership; today this would be considered very young but not at his time. John von Neumann was 31 years when he was proposed for corresponding member in 1934. Eight physicist and engineer members of the Academy nominated von Neumann and seven of the eight had been elected when they were of the age between 28 and 33 years. Von Neumann was not elected then or later, and this can be ascribed to anti-Semitism rather than to his young age. The Hungarian Academy of Sciences hardly elected

© Springer Nature Switzerland AG 2020
I. Hargittai, *Mosaic of a Scientific Life*, https://doi.org/10.1007/978-3-030-34766-6_41

Jewish scientists or scientists of Jewish origin (those who had converted or whose parents had already converted) during the 25 years of the Horthy regime.

Szent-Györgyi condemned the way the Academy was operating between the two world wars. In a letter of November 30, 1945, he wrote[1]: "... I deeply regret that the Academy refused the institution of any reform that could have renewed its operations. Without a renewal, the Academy could not be a deserving representative of the ideals of Széchenyi.[2] The Academy to a great extent is responsible for our national tragedy."

These are heavy words from 1945, but, regretfully, the Hungarian Academy of Sciences still has not faced its past. In 2016, an account, "History of the Hungarian Academy of Sciences," appeared on its official website.[3] Rather than remedying the situation, this document aggravated it. There was no mention in it of the anti-Semitic discrimination in scientific life during the Horthy era and no mention of any losses in the Holocaust, not even that there was a Holocaust (or that there was a Second World War, for that matter). I called the attention of the Department of Communication of the Academy to this misrepresentation. The response was that the inclusion of the items whose absence I lamented would not be possible in the limited framework of the treatise. It was only when I published a critical article whose title was in English translation, "Inconvenient Truth—About the History of the Hungarian Academy of Sciences"[4]—that the segment about the Horthy era in the history of the Academy was augmented by mentioning the consequences of its anti-Semitism and the *numerus clausus*.

Beyond the discrimination in the elections to the Academy, there were numerous examples of anti-Semitism concerning scientific life in the Hungary of the Horthy era. Thousands(!) were forced to study abroad, and many more did not study who could not afford it. Looking back it may seem surprising that many chose Mussolini's fascist Italy for study, but the anti-Semitic turn there came about only in 1938. Recently, a memorial plaque was erected on the wall of the busy patio at the Department of Sociology of today's Masaryk University in Brno, Czech Republic. There, between 1928 and 1938, hundreds of Jewish Hungarians studied at the German Technical University, exiled from Hungary by the *numerus clausus* legislation. The children of the "numerus clausus exiles" expressed their gratitude to the Czech Republic for hospitality, which made the studies of their parents possible. What a contrast between Horthy's Hungary that prevented their studies and the then Czechoslovakia and fascist Italy that let them study.

Another story comes to mind. There was a nationwide, large-scale anti-Jewish pogrom, the *Kristallnacht*, on

---

[1] Letter of November 30, 1945, from Albert Szent-Györgyi to the Secretary General of the Hungarian Academy of Sciences. The letter is stored under registry number 48/1946 in the Manuscript Archive of the Library of the Hungarian Academy of Sciences.

[2] The aristocrat István Széchenyi (1791–1860) helped founding the Hungarian Academy of Sciences and suggested reforms in many areas of commerce, finance, and transportation, among others, to modernize the country.

[3] http://mta.hu/data/MTA_Tortenete_ENG.pdf (downloaded, December 30, 2017).

[4] Hargittai I., *Magyar Tudomány* 2018, 179/3, 435–442.

Kindertransport memorials (by Flor Kent) at the Liverpool Street Station in London (left) and at Westbahnhof in Vienna (right). Photographs by the author.

November 9–10, 1938, in Germany, which by then had annexed Austria. Immediately afterward, 10,000 children were saved by the British, bringing them from Germany (and Austria) to the UK. This was the *Kindertransport*. This action has been commemorated not only in London where they were received, but also in Vienna, Berlin, and elsewhere from where these children were to be saved.[5]

As for Hungary, we have no information about the extent of loss of scientists in the Hungarian Holocaust that included the deportations to Auschwitz, when Nicholas Horthy was still the Governor, and the murders committed by the Hungarian Arrow Cross during the last months of the war in Budapest. The names include Gedeon Richter, Imre Bródy, Lajos Steiner, Nándor Mauthner, Frigyes Fellner, but these are only the better-known individuals. There were few members of the Science Academy among the victims, but only because of its discriminatory election practices. There has been no attempt of taking account of the losses of Hungarian scientific life due to the Horthy regime and the Holocaust.

[5] Two of my future friends were among those children. Alfred Bader (1924–2018), in part of Hungarian origin, was a Canadian-American chemist, art collector, and philanthropist. He grew up in Vienna. He accompanied me when I went to see for the first time the school in Vienna where our lager (camp) was in 1944–1945. The other, Arno Penzias (1933–), American physicist, received the Nobel Prize in Physics in 1978 for the discovery of the background microwave radiation. He wrote the Preface for one of our books.

Admired and Despised

Top: Edward Teller and the author in 1996 in the Tellers' home in Stanford, California (photograph by Magdolna Hargittai). Bottom: The Teller biography.

© Springer Nature Switzerland AG 2020   153
I. Hargittai, *Mosaic of a Scientific Life*, https://doi.org/10.1007/978-3-030-34766-6_42

Edward Teller (1908–2003) was a player on a world scale. Magdi and I visited the Tellers in their Stanford, California, home in 1996. For me, this meeting was one of the most remarkable ones among my hundreds of meetings with famous scientists. At the start, the atmosphere was unfriendly. There was then a turn for the better. Many have ascribed Teller's anti-communism to his childhood memory of the communist dictatorship in 1919 in Hungary. This always upset Teller because at the time he was only 11 years old and they did not even spend much of the time in Hungary. Letting a childhood memory shape his political views would have meant rather superficial approach, but this was not the case.

When Teller understood that I knew about the influence of Arthur Koestler's book *Darkness at Noon* on his political views, this changed his mood. The book shows the Stalinist terror through the trials of a veteran Bolshevik. Had this book been the principal influence shaping Teller's views, this might place them onto ideological rather than political foundations. This is not quite what I think happened. To me, Teller's anti-communism was rather political than ideological. Accordingly, I would call what Teller represented more anti-Soviet than anti-communist. He wanted to protect the Free World and most of all the USA first from Nazi Germany and later from the even more menacing Soviet Union.

However pretentious it may sound, I think that my Teller biography[1] has been the most objective among his biographies, and the reviewers have also found it balanced. This is worth mentioning because Teller has been admired or despised in the USA. Recently, my impression has been that he may become gradually forgotten, which Teller might have found even worse than being despised. For a couple of years, I immersed myself in Teller's life to a degree that it was almost indiscretion.

He had many admirable traits and many condemnable. He had will power, which helped him working on friendships during the first decades of his life. He started with no friends among his contemporaries and then became popular. In the second half of his life, he lost his friends as a consequence of a self-destroying crusade for winning his battles and sought the company of military and political leaders. He assisted in revoking the security clearance of J. Robert Oppenheimer, the "father of the American atomic bomb." Oppenheimer had opposed the development of the hydrogen bomb, which Teller desperately wanted. Teller was right in advocating the need for the American hydrogen bomb, but his action against Oppenheimer was more revenge than necessity. Most of the American physicists opposed the development of the hydrogen bomb, because they did not feel that the Soviets were capable of producing one—in this, they were wrong and Teller was right. He was also instrumental in finding the way to build this horrible weapon. His fear of the Soviet nuclear program was based on his recognition of the acumen of Soviet physicists and of the capability of a totalitarian state focusing on and accomplishing extraordinary projects in spite of its underdeveloped infrastructure.

The following episode in Teller's life is most revealing of his will power. He was 20 years old when in a Munich tram accident he lost one of his feet and had to live with prosthesis for the rest of his life. It was during the recuperation following one of his surgeries that he noticed how the painkillers weakened his capacity of thinking—his favorite pastime. He decided that he would not take painkillers anymore. Rather, he would just go through the motion of taking the pill without actually taking it. He would swallow as if he had taken the pill, drink a little water, and think as if he had taken the pill, although he did not. This approach worked; his will won over his pain. Most people who came across Teller did not even realize that he was wearing prosthesis and he did everything in his life as if he did not wear one. Toward the end of his life when he was in a wheelchair and could not see anymore, he complained that the only thing in his body that was still operating properly was his prosthesis.

Teller let his pain of losing his friends be manifested only on the rarest of occasions, but it was there, however deeply buried. He left his anti-Semitic home country when he was 18; then, at 25 years, he had to leave Germany where both for its language and its culture he felt himself very much at home. His third "exile" came in 1954 when the role he played in the Oppenheimer hearing forced him out of the community of his fellow physicists. This was his most painful "emigration."

[1] István Hargittai, *Judging Edward Teller: A Closer Look at One of the Most Influential Scientists of the Twentieth Century* (Amherst, New York, 2010).

Jeffrey L. Levine, deputy chief of mission of the Budapest US Embassy, and the author on January 15, 2008, at the unveiling of the Teller memorial plaque in Budapest, 3 Szalay Street (by unknown photographer).

Teller's *Memoirs*[2] provide an enjoyable read, but they are not always realistic about his deeds. It is not necessarily misleading on purpose; rather, his remembering some of the events must have undergone a metamorphosis. He described them as he wished they had happened rather than the way they actually did. In addition to factual errors, there are stories that are narrated so vaguely that may be interpreted in more than one way. Intimation was a technique he used with great skill not only in his *Memoirs* but also in debates.

Here is an example of the way how he used intimation. Teller greatly admired his mentor in his doctoral work, Werner Heisenberg (1901–1976), one of the greatest physicists of the twentieth century. Heisenberg was in charge of the atomic bomb project in Nazi Germany. The project failed and after the war Heisenberg claimed moral victory over the American physicists of the Manhattan Project who did produce the bomb. Teller, who for many years, including the war years, did not meet with Heisenberg, and could not know much about Heisenberg's activities and demeanor during the Nazi years, accepted Heisenberg's claims. Teller referred to what Heisenberg had said that he would die with peace in his soul. For Teller this sufficed as proof that Heisenberg did not want to have the German project succeed. This was a shaky argument, and not unique, as such, in Teller's practice.

Teller's performances in debates—and he was an outstanding debater—suggest that for him the end justified the means. This expression is apt here, because he has been compared to Machiavelli. In some cases, he argued that radioactive contamination from nuclear testing represented

---

[2] Edward Teller with Judith Shoolery, *Memoirs: A Twentieth-Century Journey in Science and Politics* (Cambridge, Massachusetts: Perseus, 2001).

Anders Bárány, former long-time secretary of the Nobel Prize Committee of Physics in fall 2015 visiting Leo Szilard's grave at the academicians' section in the Fiume Avenue National Necropolis in Budapest (photograph by the author) (Half of Szilard's ashes rest in the Budapest cemetery; the other half is buried in a cemetery in Ithaca, New York).

no danger. In other cases, he argued that the birth defects as a consequence of nuclear testing were an acceptable price to be paid for enhancing national security.

The Hungarian "Martians" stayed friends with Teller irrespective whether or not they agreed with him politically. This did not prevent Leo Szilard from having fierce public debates with Teller. Szilard has been considered the opposite of Teller politically in the oversimplifying media. Teller has been considered by many as a warmonger and Szilard as an advocate of peaceful coexistence with the Soviet Union. There was, however, one crucial question in which the two did agree. That was about the necessity of the development of

the American hydrogen bomb. When I first read about this, that is, that Szilard considered the development of the American hydrogen bomb unavoidable, I was surprised. I decided to look for proof that this was the case indeed. I found the proof in 2004, in my encounter with the famous professor of molecular biology of Harvard University, Matthew S. Meselson (1930–). In 1954, Meselson was a young researcher in California and he was the guide for the visiting Leo Szilard who arrived for a lecture. Szilard considered this particular lecture so important that he practiced it in front of Meselson as a one-person audience. Thus, Meselson heard it from Szilard not once, but twice that Szilard supported

Teller's struggle for the development of the hydrogen bomb. To this day, many have condemned Teller for influencing the USA to develop the hydrogen bomb. This is, of course, an exaggeration of Teller's importance, but this is beside the point. This view is sadly wrong because even while the American physicists were still debating whether to develop the hydrogen bomb or not, the Soviets had already embarked on working on it. The Americans did not have proper intelligence to know about the Soviet nuclear project whereas the Soviet intelligence followed the American progress in minute detail. It is not a pleasant line of thought to consider what might have happened had Stalin acquired the hydrogen bomb first.

Lev V. Vilkov and the author in the 1970s at the Moscow Kreml (by unknown photographer).

In my career in science, Lev V. Vilkov (1931–2010) was the only person whom I could consider directly as a supervisor. This lasted the academic year, 1964/65, when I was preparing my master's thesis. So, it was for about 9 months, but it was also a very intensive interaction. This was my last year at Lomonosov University; there were no subjects or examinations; it was research, day and night, the determination of a molecular structure. My interaction with Vilkov continued until the end of his life; only its character changed. It became a partnership of friends and it did not diminish during the political changes in 1989–1990 in Hungary when many of my colleagues severed their former Soviet connections. In 1990, the State Secretary for Education

handed me the diploma of university professor, signed by the President of the Republic. There was a small reception on this occasion and the State Secretary asked me a leading question concerning the comparison of my experiences in American and Soviet laboratories. At the time former graduates of Soviet universities tried to dissociate themselves from their past and on one occasion such a former graduate even denied that he could speak Russian. Amnesia was in vogue. So, on that occasion, I told the State Secretary that the training I received for scientific research in Moscow was of the highest level and there was no influence of ideology or politics within the laboratory. In this I saw no difference between the Soviet and American laboratories. If there was a difference, it was in how the two places handled the international literature. For the Americans it was virtually delivered to their desk, but few followed it. My Moscow colleagues read it religiously although it was very difficult for them to gain access to it. There was some defiance in my response to the Secretary's question; the reality was a little more complicated. Even at the time of my studies, ideological considerations hindered the penetration of some theoretical advances into our curriculum although these limitations were much diminished as compared with earlier times.

the same room and a darkroom was in it as well. There were then his students, myself included, working on their theses. Whenever we wanted to talk with our mentor, we had to wait our turn, which had the advantage of learning from his discussions with others. There were two other similar rooms of the Electron Diffraction Laboratory, each occupied by a professor-ranking person like Vilkov and their respective groups. The three groups used to gather for joint seminars where the talks were given by associates of the Laboratory, students, and occasionally by international visitors. The

Lev V. Vilkov and the author in 1965 in Moscow in the break of a soccer game (by unknown photographer).

There was a sizzling life in the Moscow laboratory, facilitated by the crowded conditions. The new campus of Lomonosov University was inaugurated in 1953 and there were spacious arrangements. By the time I was a student there, space had become scarce; even some of the restrooms had been turned into laboratories. Vilkov's study was the headquarters of his group with fewer desks than the number of his associates. The electron diffraction apparatus was in

sometimes fierce debates were especially instructive. One of the leaders of the three groups, Nikolai R., was the most active in these discussions. He liked the debates for the sake of debating and he was very good in them. He could argue forcefully for an issue until somebody would point out an elementary error in the initial conditions. Nikolai R. then stopped, but he only needed a few moments to regroup, and repeated his argument now with the opposite conclusion. I did not find his style attractive then and I do not find it attractive now, but that he could argue back and force showed his gift as debater. Now I know that in America, one can learn such arts as part of the curriculum and there are circles for practicing debate—a very useful exercise.

Vilkov participated in the debates but I could have hardly imagined him reversing his point of view with such an ease as Nikolai R. did. Vilkov followed the literature and expected his students to do the same. He was well versed in it in spite of his weak skills in the English language. For his generation it was not encouraged to learn English. He valued international interactions as did the two other leaders. It was advantageous that none of the three groups was involved in classified research as some others in our department were, as witnessed by meticulously locked rooms in our hall.

Upon the defense of my master's thesis in 1965, Vilkov suggested to me to stay on for doctoral studies. In 3 years, I could have become a PhD. The University prepared a formal recommendation for the Hungarian authorities. I did not have a job yet in Budapest and it seemed a straightforward continuation of my studies. I had to visit the Secretary General of the Hungarian Academy of Sciences, Tibor Erdey-Grúz, who was supposed to give the final approval. However, the encounter was far from superficial. Erdey-Grúz appeared genuinely interested in the research I was planning to do and asked what turned out to be a crucial consideration: "Do we want to have this area of science in Hungary?" When I said yes, he made the decision to start it right away. I soon overcame my initial disappointment. It was not easy, and my PhD equivalent degree took longer and more work,

but I became independent and could contribute to my science creatively. The final outcome justified Erdey-Grúz's decision, which at the time seemed to me quite haphazard as it probably was.

Around 1990, at the time of the political changes when many in Moscow were starving, and this is no exaggeration, we sent a suitcase of food to the Vilkov family. We were just having an American Nobel laureate visiting with us. He was surprised because he did not think we were much better off than the Russians—we were, but from his point of view there was not much of a difference. He understood though that we were more sensitive to the Russians' plight and had more empathy for the Russians being closer to their trials. A suitcase of food was no long-term solution. What saved Vilkov and many others in this situation was a Soros Stipend they received to help them over this crisis.[1] This Stipend was meant to help Russian scientists do quality work and stay in Russia.

In spite of the political changes, my interactions with Vilkov even strengthened in the 1990s as I moved from an academy position to a university position. I became eligible to have doctoral students and my first doctoral student was a graduate from Vilkov's laboratory. I visited Vilkov's former group recently. What used to be his room is no longer crowded and the laboratory life is no longer so sizzling as it used to be. During the past decades, the physical conditions have further deteriorated, but that was not the decisive factor. What was missing was Vilkov's stimulating demeanor, the atmosphere of his intellectual curiosity, and his contagious creative excitement.

---

[1] George Soros (1930–) is a Hungarian-born American financier–investor and philanthropist having supported democratic changes in Hungary and the rest of Eastern Europe.

A Glimpse into His Thinking

James D. Watson and the author in 2000. Top: In Watson's office at the Cold Spring Harbor Laboratory. Bottom: In the Hargittais' home in Budapest (photographs by Magdolna Hargittai).

James D. Watson (1928–) has secured his place in the annals of science history as the co-discoverer of the double-helical structure of DNA. He and Francis Crick suggested a structure that became popular overnight because of its beauty in addition to its scientific significance. They recognized the possible copying mechanism and the hereditary properties of DNA. Watson's achievements include his bestseller *The Double Helix*, which brought scientific research into human proximity for a broad readership.

For me, personally, he made me a gift by letting me glimpse into his thinking. We first met early 2000 when Magdi and I visited him at the Cold Spring Harbor Laboratory to record a conversation with him. The start was unpleasant as I was receiving curt responses and his inner world appeared closed before me. Then, there was a point when everything changed and the encounter turned into a sizzling intellectual adventure. We met again a few months later, in summer 2000, when Watson and his wife, Elizabeth, visited us in Budapest. In the focus of their crowded schedule was a meeting with Hungarian intellectuals that we organized in our home at his request.

Our next meeting was again at the Cold Spring Harbor Laboratory where we spent 3 months in spring 2002 at the personal invitation of the Watsons. The purpose of the stay was to work on my semiautobiographical book *Our Lives* (see the Árpád Göncz chapter). There were other planned and unplanned meetings. In 2003, we attended the 50th anniversary celebrations of the discovery of the double helix in Cambridge. In 2004, while visiting Matthew Meselson in Woods Hole, the Watsons stopped by. Almost every time we were in the USA, we went to see the Watsons either in Cold Spring Harbor or in their Manhattan home.

In 2007, Watson gave an interview to one of his former associates who worked for a newspaper. Unbeknownst to Watson, she left on her recorder even when they were having what he thought was a private conversation. She then quoted Watson's comments about the hopelessness of Africa and the Africans that could be considered racist. There was a devastating uproar upon the publication of the report. Virtually everybody who was close to Watson disowned him, including the Cold Spring Harbor Laboratory, which had become a world leader in research due to Watson's efforts. The Laboratory did not allow him to speak with the press for a while and forced him into retirement. They assigned a publicist to him and for all practical purposes, placed the fiercely independent Watson under guardianship.

I knew that Watson was not a racist, but had an inclination to make shocking statements. We did not distance ourselves from him and to a small part eased his isolation. He characterized his situation with the word "sordid," a word he used for it repeatedly. Then, in 2008, he was allowed to be interviewed as arranged by his publicist, and Watson asked me to be present. It was a sad occasion. The reporter appeared to be ignorant of Watson's achievements and persona. She asked all the requisite questions, without any follow-up. Her full attention was on keyboarding Watson's responses into her computer. It seemed to me that the Laboratory had no confidence in Watson and tried to further confine his independence rather than letting him to be himself. This was the lowest point.[1]

In subsequent years, the situation gradually improved. The atmosphere of our meetings—by now invariably in their Manhattan home—returned to the old pleasant state. In the meantime, the Laboratory had realized that Watson's ability of generating support for it remained unchanged and it kept benefiting from it. Watson returned to the limelight when he let his Nobel medal be sold at an auction. A Russian billionaire bought it only to have it returned to Watson. There was a ceremony for this event in Moscow and the President of the Russian Academy of Sciences handed back the medal to Watson at the headquarters of the Academy.

---

[1] Added in summer 2019: Unfortunately, in spite of his apologies for his statement in 2007, Watson reiterated his views about the interrelationship between intelligence and genetics in mid-2018, and it sounded awful. The Cold Spring Harbor Laboratory revoked all his honorary titles. Watson is in no position to respond. He has been incapacitated due to an automobile accident.

James D. Watson and the author in 2010 in the Watsons' home in Manhattan (photograph by Magdolna Hargittai).

Watson's public appearances have become rare. He likes to talk to young people, as he did in Moscow. He gives advice to them about how to achieve success in science. For progress in research, he considers the understanding of how the brain works and how it stores information to be of the greatest interest. In this, the illness (bipolar disorder) of one of his two sons plays a role.

Watson has always been interested in the roots of his own success. This is not unique among Nobel laureates, but he has posed more realistic questions than most. One of his puzzles is how could he, not being a truly great scientist, climb to the summit in science and stay there for decades. Of course, we should not confuse "great scientist" and "scientist genius." Risking oversimplification, the former knows everything in a field; the latter has the ability to recognize contexts others might have not. There is no doubt that Watson is a scientist genius.

He has always been controversial and, especially when he was director, he hurt people. I did not have personal experience in this but have heard stories that were consistent and sounded realistic. There were stories characteristic of a sulky child, but more serious ones, too. He encouraged rivalry between sections, between research groups, and between individual scientists in the Laboratory. This contributed to creating a hostile environment. The impact of his persona on the Laboratory was so overwhelming that no other strong individual would be willing to take up the leading position when Watson's succession was on the agenda. He has been in charge, undisputed, whether he was director, president, or chancellor, or even having no formal position.

Luckily, Crick and Watson eminently complemented each other in their joint work. There are similarities and differences between them, and here I single out only one of their differences. They both were curious and ambitious. However, Watson's ambition was stronger than his curiosity whereas Crick's curiosity exceeded his ambitions. The discovery of the double helix was less determining for the rest of Crick's career than for Watson's. Watson has built a cult in which he and DNA, and not just its structure, could be worshipped interchangeably. Crick left DNA, and moved elsewhere facing a greater yet challenge than before, the research of consciousness. Watson is more the public's scientist; Crick was a scientists' scientist.

As for me personally, Watson was very helpful and I have felt the positive impact of his actions, starting with his support for my *Our Lives*. All this I ascribe to his appreciation of Hungarian intellectuals in general, and such people in particular as Leo Szilard and George Klein. My two recorded conversations with Watson, augmented by one Magdi recorded with him, served as the basis for a book in which I attempted to present him from the viewpoint of the scientific community.[2]

[2] István Hargittai, *The DNA Doctor: Conversations with James D. Watson* (Singapore: World Scientific, 2007).

Beautiful Math

Richárd Wiegandt and the author on the "Day of the Book" in 2011 at Vörösmarty Square in Budapest (photograph by Magdolna Hargittai).

Richárd Wiegandt (1932–) taught mathematics between 1955 and 1961 at the Orosháza Táncsics Gimnázium, which I attended in 1957–1959, that is, from the middle of my sophomore year through graduation. Wiegandt was my favorite teacher and math was my favorite subject. When I joined the school, he had already been teaching my class for a year and a half, yet he treated me as if I had been in the class from the beginning. This was the best way to handle the situation. I learned only later that he had been applying for years for postgraduate studies, unsuccessfully. In this light, it is even more remarkable that he was always patient even with pupils for whom math remained hopelessly inaccessible.

My love of math was selective. I enjoyed the derivation of the expressions of volume for polyhedra, for example, the truncated pyramid. I doubt that it was part of the official curriculum, but Wiegandt showed these derivations though never asked the students to repeat them during recitation. My favorite was the exponential equation in which the unknown is in the exponent. Of course, our curriculum included only the simplest kind. Wiegandt used to write the equation to be solved on the blackboard and only then did he ask one of us to come forward and solve it. The ensuing few seconds sufficed for me to solve the equation by a trial and error technique. By the time a student was selected to solve the

© Springer Nature Switzerland AG 2020
I. Hargittai, *Mosaic of a Scientific Life*, https://doi.org/10.1007/978-3-030-34766-6_45

equation, I could whisper the solution. Of course, not every student needed such assistance, but it was appreciated. If our teacher was surprised that even the weakest students in math produced the solution instantly, he did not show it. He must have found the problems so easy that in his eyes they must have been child's play.

For Wiegandt, the students were partners in his teaching and in this he was different from most of the other teachers. He was polite and did not scold us—there was no need for it, because he captivated the attention of the students, even those who did not really understand what he was talking about. He was a true pedagogue who educated us, not only taught, and by example rather than by words.

I mention a few other teachers whom I liked too in the same school. György Fancsovits taught history, which was not an easy assignment, especially when we were approaching the recent past and he could not deviate from the official curriculum. Yet he wanted at least some of us to receive a broader perspective. He used to invite a group of students to his home for further discussion. His wife, who also taught in our school, tolerated kindly our noisy presence in their tiny apartment.

came around and so did some of the teachers. Kazár developed some informal chemistry class and he talked about topics that he thought might interest us. For example, he introduced us to the phenomenon of isomerism. Suppose a molecule contains four carbon atoms. These four carbons may form a chain or may form a chain of three carbon atoms with the fourth linked to the middle carbon in the chain. The four-carbon molecule may have two possibilities of the carbon arrangements.[1] The larger the number of carbon atoms in the molecule, the larger the number of possibilities of their arrangements, and the number of possibilities increases rapidly. This is the origin of the great variability of carbon compounds. It is a captivating task to find all possible isomers for a given number of carbon atoms. In the official curriculum, very little time was assigned to this question whereas playing with this was entertaining and instructing.

Returning to Wiegandt, in 1961 (I had by then left the school), he was accepted for postgraduate studies. László Fuchs (1924–) was his mentor, an internationally renowned mathematician who later had a distinguished career in the USA. Wiegandt earned his higher qualifications in

The Orosháza Táncsics Mihály Gimnázium (photograph by Mihály Varjú; courtesy of János Blahó).

József Kazár was another teacher with whom I had positive experience. This happened in the early months of 1957 soon following the suppression of the 1956 Revolution when there was still general strike in Hungary. There were no instructions in the Gimnázium, but it was open and students

---

[1] This was isomerism of connectivity: Two molecules of the same composition having different orders in which their atoms are connected. This is a different kind of isomerism from the one presented in the Degas chapter.

mathematics, became an associate of the Institute of Mathematics of the Hungarian Academy of Sciences, and gained fame as visiting scientist in a number of countries on five continents.

I had a rather special teacher in my last year, the eighth grade, of general school in Orosháza, Mrs. Horváth. During that year, we were saving up funds for a class excursion following graduation. This was an all-boys class of very diverse composition, including a number of gipsy students. There were no exceptions, all of us made the trip of 2 weeks to the Bükk Mountains, and everything was financed by our earnings as a collective. If there ever was a preparation for grownup life, this was.

Initially I was hesitant whether Orosháza should figure in this volume, because of my painful memories of this town. This is where we were deported from in 1944 and this is where we returned to in 1945, only to find our home ransacked. This is where the officialdom tried to severely curtail my possibilities when one of my schools wanted to reward me for good performance. This is where I was initially declined the possibility of studying in the local gimnázium, and this is where a negative characterization was prepared attempting to prevent my acceptance as a university student. However, it was also here where I had some good teachers and good friends, and where I am still maintaining good interaction with my Alma Mater. I was seeking some further positive experience that I could mention in connection with this town, and I finally succeeded. It is based on narrative by family members, but it lives on in me as if it had been in my own memory—something that happened in spring 1944. It was when we were already limited in various activities but before we were forced into the ghetto. In good weather, our windows were open in the mornings, and every morning someone threw a sack of fresh rolls into our home through the open windows. It was Mrs. Lövei of the local Lövei Bakery.

He Taught Me Symmetry

Eugene P. Wigner and the author in 1969 on the campus of the University of Texas at Austin (by unknown photographer).

© Springer Nature Switzerland AG 2020
I. Hargittai, *Mosaic of a Scientific Life*, https://doi.org/10.1007/978-3-030-34766-6_46

I first interacted with the Hungarian-American scientist Eugene P. Wigner (1902–1995) in 1964 through correspondence. He was born in Budapest and died in Princeton. He was awarded the Nobel Prize in Physics in 1963 for the application of symmetry principles in nuclear physics. The community of physicists learned about Wigner's distinction with great satisfaction. In Hungary, there used to be ambivalence about how to consider the successes of expatriate Hungarians. Using the occasion of the Nobel Prize, the literary magazine *Élet és Irodalom* (Life and Literature) printed an article about Wigner's high school years in Budapest. In response, Wigner sent the magazine one of his papers, which the magazine published in fall 1964 in Hungarian translation. This philosophical paper from 1950 was about the limits of science. The magazine added Wigner's note that he would be curious of how people in Hungary think about the limits of science.

In fall 1964, I was working on my diploma work (master's thesis) in physical chemistry at Lomonosov University. I had a subscription for *Élet és Irodalom*. When I read Wigner's paper and his note, I knew that I had to respond. In hindsight, this was peculiar, because until then I had never had the urge to express myself in such a way and nothing had appeared by me in print. I sent my response to the magazine, which published it in a few weeks' time. In a few more weeks, I received a large envelope from Wigner with a letter and a bunch of reprints. Wigner's letter was factual; he elaborated what he agreed with me about and also—in his most polite way—what he did not agree with me about. This was the beginning of our on and off life-long correspondence. In person, we met only once.

In 1969, I was a visiting research associate at the Physics Department, University of Texas at Austin (see the Bastiansen chapter). The Physics Department at the time aggressively sought out famous scientists to hire or at least to have them for a visit. This is how I met Michael Polanyi (see a separate chapter) and this is also how I met Wigner. When I heard about his forthcoming visit, I asked the new chairman of the department, a Dutch physicist, to let me see Wigner when he comes. The chairman told me that Wigner's time was so expensive that they could not afford to spend any of it with a visiting associate. However, he let me leave a note for Wigner, saying that I was around and I gave him information about my office and lab.

From the second day of his visit, Wigner came to my office every morning at 8 am, spent about 50 minutes with me and left so that by 9 am he could report in the chairman's office. Thus, he was spending his "private" time with me at no loss to the department. He had a crowded schedule indeed and he gave several lectures. One of them was about his pet project, civil defense. He showed a photograph of the Budapest metro as an example of efficient means of civil defense.

Spending time with Wigner was a unique experience. The world's best qualified person was introducing me to the applications of the symmetry concept in a series of personal tutorials. He stressed the universality of the concept and that we should rid ourselves of the dogmatic division of science into disciplines like research councils and our schools do.

We talked about other things as well, such as poetry. The Hungarian Mihály Vörösmarty was his favorite from his high school years. There was a Vörösmarty poem that inspired him and a few lines from it appeared in several of his presentations, "The world won't last forever/But while it lives and while it lasts/It builds or rents/But never rests."[1] According to Wigner, these lines expressed his own desire to have purpose, act upon it, and as a result, leave something behind. Wigner was interested in later Hungarian poetry and the martyr poet, Miklós Radnóti, especially captured his attention. I happened to have with me a *Radnóti Complete* and I gave it to him.

We talked about Wigner's time in Hungary. He was most apprehensive about the short period he lived there and his brief visits there between the two world wars. However, he spoke with appreciation about his high school, the *Lutheran Gimnázium*, which he attended between 1912 and 1920. He singled out his math teacher, László Rátz, and his fellow student, one year his junior, John von Neumann, as lasting influence on him. Both Wigner and von Neumann attended the Lutheran high school as Jewish pupils. Many Jewish pupils attended the Lutheran high school, one of the best in Budapest, although the tuition for Jewish pupils was multiple times that of the Lutheran students.[2]

Wigner spoke bitterly about what he experienced during his visits to Hungary in the 1920s and 1930s. Whenever he felt homesick, it sufficed for him to remember those visits, and the feeling evaporated. The officials in the offices treated the ordinary citizens as beggars rather than clients. At the same time, he wanted to belittle his negative experiences. Laura Fermi, Enrico Fermi's wife, wrote about this in her book *Illustrious Immigrants*. She knew about the everyday beatings of the Jewish students by nationalistic students in Hungarian universities at Wigner's time. When she asked him about this, he did not remember any such incidents. When pressed further, and he admitted that, yes, they also beat him, but, added hastily, it did not hurt much. When I

---

[1] See, e.g., *The Collected Works of Eugene P. Wigner, Volume VII: Historical and Biographical Reflections and Syntheses*, annotated and edited by Jagdish Mehra (Berlin, Heidelberg, New York: Springer-Verlag, 2001); quotations: pp. 319; 392; and 463.

[2] The tuition for Calvinist, Catholic, and Jewish pupils were twice, three times, and five times, respectively, the tuition of the Lutheran pupils. Wigner's family converted during the last year of his high school studies.

asked him about this, he did not remember such experience. He added though, had he remembered, he would not tell me.

Nowadays, in the 2010s, we hear about the Horthy regime's attempts of repatriating Hungarian emigrants. However, such attempts did not cover Wigner or von Neumann. The anti-Semitic Horthy regime did not consider them Hungarian. In the mid-1930s, there was an opening for a professorship of physics at the University of Szeged. The renowned physics professor Rudolf Ortvay mentioned this possibility to Wigner. Wigner wrote in response to Ortvay that "... The predominance of these non-scientific considerations would exclude my chances for this job, and the predominance of these very considerations makes this job for me hardly desirable."[3]

It was legendary how keenly Wigner watched out for being "reasonable" yet my impression was that his interest in Hungarian politics, even in the utterances of insignificant Hungarian politicians, was unreasonable. Politically he was very conservative. At the time of our meeting, in 1969, it was the first year of the Nixon Administration, but Lyndon B. Johnson's "War on Poverty" was still fresh in memory. However, according to Wigner, poverty did not exist in America. He wanted to prove it to me and invited me to visit him in his home. He wanted to show me around in the neighborhood so that I see it with my own eyes. Only later did I learn that Mercer County where he lived was among the highest income counties in the USA.

Forty years later, István Hargittai with Wigner's bust in 2009 on the campus of the Budapest University of Technology and Economics (courtesy of Eszter Hargittai).

---

[3] Wigner's letter of January 13, 1936, to Ortvay. Identification: K785/139; Archive of Manuscripts of the Library of the Hungarian Academy of Sciences. My translation.

Epilogue

The measure of success is
the achievements under the existing conditions rather than
what could have been achieved under ideal conditions
(*István Hargittai*)

Some time ago Géza Komoróczy[1] asked me for the one word I would use to characterize my life, and I had to respond at once, without deliberation. My response was, "struggle" ("küzdelmes," in Hungarian). When I gave myself more time to think about it, I still could not come up with a better answer. I realize that the English "struggle" sounds a little more combative than my choice in Hungarian, but still it comes closest. When I think about the beginning of my life, the rest is almost unbelievable. I did not think about it much, except lately, that my country condemned me to death when I was not yet three years old, and then let me back, however reluctantly. This ambivalence I have felt throughout my entire life, regardless of which political system was in power.

I should have stopped thinking about this long ago, but circumstances do not let me. I see that the regime that sent me to perish has become the model of the past for the present leadership. What saves me is that it is possible to carry on living while ignoring all that I could not change. These are heavy words, but my optimism should shine through them. I have asked my children when the time comes to remember me for my humor rather than any gloom.

As I was compiling this volume, and building the mosaic by always adding one piece at a time, the list of names left to consider was not shrinking. Alas, I had to stop at some point lest it become endless. Nonetheless, the names that stay within me, and the people behind the names, leave me feeling rich and grateful.

---

[1] Professor Géza Komoróczy, PhD (1937–), Hebraist, Assyriologist, author, and historian, to me epitomizes the scientist to whom it is possible to turn for reliable information, who though non-Jewish can even in Jewish history outperform the most knowledgeable Jewish scholars. He epitomizes to me the archetypal academician, but who cannot be a member of the Science Academy lacking the prerequisite higher scientific degree for which he has declined repeatedly to apply.

© Springer Nature Switzerland AG 2020
I. Hargittai, *Mosaic of a Scientific Life*, https://doi.org/10.1007/978-3-030-34766-6

Culture and Art of Scientific Discoveries

*A Selection of
István Hargittai's Writings*

Balazs Hargittai
Editor

 Springer

The cover of my latest book, a selection from my non-technical writings, compiled by Balazs Hargittai

As Agnes Heller alluded to in the Foreword, my interests have shifted from molecular structure to more general questions of science, such as the nature of scientific discovery. Our son, Balazs, also a chemistry professor, has recently compiled a selection of my non-technical papers in a large volume.[2]

Looking back, during my career I was fortunate to be doing what I liked best, most of the time. I realize that there seems to be a contradiction between my "struggle" mentioned above and this last statement. If there is a paradox in this, I like paradoxes.

---

[2] Balazs Hargittai, Ed., *Culture and Art of Scientific Discoveries: A Selection of István Hargittai's Writings* (Cham, Switzerland: Springer Nature Switzerland, 2019).

Index

Numerals
1918, 134
1919, 16, 134, 154
1933, 56, 108, 110, 134, 151
1956, 5, 10, 14, 17, 36, 47, 67, 71, 120, 122, 143, 144, 168
1968, 1, 2, 8–10, 48, 98

A
Abrikosov, A., 30
Abstract art, 63
American Chemical, 121
American Chemical Society, 145
American Physical Society, 102
Amino acids, 10, 11, 27
Aminoff Prize, 85
Anthem, 94
Anti-communism, 154
Anti-Jewish legislation, 16, 94
Anti-Semitism, 34, 70, 74, 110, 111, 149, 150
Applewhite, E.J., 59, 60
Archives, 79, 91, 105, 135, 139, 140, 143, 144
Arrow Cross (Hungarian Nazis), 24, 94, 122, 123, 151
Artistic freedom, 80
Arzamas-16, 41, 140
Atomic bomb, 134, 154, 155
Atomic Energy Commission (USA), 145
Auschwitz, 26, 35, 36, 46, 54, 70, 94, 108, 151
Auschwitz Protocol (Vrba and Wetzler), 70
Austrian lagers (labor camps), 26, 46
Autocratic regimes, 17, 48
Avant-garde art, 67
Awakenings (Sacks), 49

B
Bader, A., 32, 151
Balazs, E., 6, 7, 83
Balázs, J., 66
Balogh, T., 135
Bárány, A., 156
Bard College, 5
Bartók, B., 128
Barton, D., 120
Bastiansen, O., 1–3, 27, 172
Bauhaus, 78
Bécsi Avenue, 66
Ben-Gurion University, 117
Berkeley (University of California), 35
Bernal, J.D., 11, 101, 102, 126
Biomatrix (company), 8

Birkbeck College, 101, 126
Bitó, L., 5–8, 83
Black, J.W., 30, 111
Bohr, N., 134
Bonner, E., 140
Boston University, 2
Bragg, W.L., 101
Brenner, S., 9, 10
Bródy, I., 151
Brown, L.M., 35
Brunvoll, J., 2, 3, 107
Buckminsterfullerene, 58, 59, 84, 137
Budapest Scientific (Hargittai és Hargittai), 43
Budapest University of Technology and Economics, 36, 119, 121, 122, 173
Buried Glory (Hargittai), 41, 139, 140

C
California Institute of Technology, 10, 110
Caltech (California Institute of Technology), 110, 111
Calvin, M., 134
Cambridge, 10, 11, 36, 61, 101, 120, 149, 155, 164
Camon, F., 110
Camps (lagers), 26, 46, 94
Candid Science (Hargittai, Hargittai, and Hargittai), 34, 38, 54, 117, 144–147
Centenary of the Nobel Prize, 119
Central European University (CEU), 41
Chain of innovation, 8
Chemical Heritage, 31, 32, 145
Chemical Intelligencer, The, 78, 79, 123, 137, 138, 144, 145
Chulabhorn Research Institute, 31
Chulabhorn (Royal Princess), 31
Ciechanover, A., 54–56
Civil defense, 172
Cold Spring Harbor Laboratory, 45, 163, 164
Columbia University, 5–8, 31
Complementary Kitaigorodsky (Orosz), 125
Compositio Mathematica, 26
Computers and Mathematics with Applications, 62
Conformism, 69, 70
Conversations with Primo Levi (Camon), 110
Cooperations, 3, 83, 107, 108, 134
Corruption, 42
Courant Institute, 93–95
Courant, R., 93, 94
Courier, 20
Coxeter, D., 25
Crick, F., 9–12, 119, 164, 165
Crick, O., 10, 11

Csanády, A., 137
Csocsán, G., 14
Csonkás, M., 13–14, 115
Culture and Art of Scientific Discoveries (Hargittai and Hargittai), 7, 21
Culture of Chemistry (Hargittai and Hargittai), 21, 78, 79, 123, 138

D
Dalos, G., 15–17
Damadian, R., 90
Darkness at Noon (Koestler), 73, 154
Data documentation/banks, 24
Debates, 15, 29, 30, 34, 35, 121, 123, 135, 145, 155–157, 160, 161
Degas, E., 20–21
Déjà vu, 16, 20, 72
Democracy, 6, 16, 29, 135
Deportation, 16, 35, 66, 151
Der Untertan (Heinrich Mann), 70
Descartes, René
Dictatorships, 5, 7, 29, 69, 70, 80, 94, 98, 103, 134, 154
Diplomats/diplomacy, 41–43, 104, 118
Discoveries/innovations
    adsorption (M. Polanyi), 133–135
    background microwave radiation (Penzias), 151
    bioenergetics (Ernster, Szent-Györgyi), 24
    buckminsterfullerene (Curl, Kroto, Smalley et al.), 59, 84
    chemical chain reactions (Semenov), 143
    combinatorical chemistry (Furka), 27, 28
    double-helix of DNA (Watson and Crick), 9, 11, 164, 165
    generalized crystallography (Mackay), 77, 85, 100–102
    genetic code (Crick, Watson), 9, 11
    hydrocarbon chemistry (Olah), 123
    immunological insight (Klein), 69
    mechanism of reactions (M. Polányi, Olah), 121, 133, 135, 143, 164
    in molecular genetics (Weissmann), 35
    MRI (Lauterbur, Mansfield), 36, 89–91
    NO in physiology (Furchgott, Moncada), 35
    in nuclear physics (Teller), 126
    particle physics (Marx, Ne'eman, Gell-Mann), 103
    photosynthesis (Calvin), 134
    for polysaccharides (Laurent), 83, 84
    purification of hyaluronic acid (Balazs), 8
    quasicrystals (Shechtman, Mackay, Steinhardt, Levine), 85, 100–102
    superacids (Gillespie, Oláh), 38
    surface chemistry (Somorjai), 36
    ubiquitin (Hershko, Ciechanover, and Rose), 54
    Vitamin C (Szent-Györgyi), 149
    in X-ray crystallography (M. Polanyi), 133
Discrimination, 94, 104, 111, 150
DNA, 9, 11, 164, 165
Doctor DNA (Hargittai), 165
Dogmas, 75, 99, 100, 102, 135, 172
Dürer, A., 102
Durston, J.H., 75

E
Eclectica (Mackay), 102
Elections to the Academy of Sciences, 150
Élet és Irodalom (Life and Literature), 46, 172
Elias, J., 138
Emerson, R.W., xiii
Emigration, 7, 34, 101, 103, 104, 134, 154
Émigrés, 103, 104, 135
Englert, F., 118

Eötvös University, 1, 27–29, 36, 97, 98, 103
Epistemology, 133
Erdey-Grúz, T., 161
Erdős, P., 26
Ernster, L., 23–24, 83, 84
Ernst, R., 91
Ertl, G., 36, 133
Escher, M.C., 126

F
Falsification of history, 46, 94
Fancsovits, G., 168
Fedezzük föl a szimmetriát (Discover Symmetry, Hargittai and Hargittai), 85
Fedin, E., 47
Fejér, L., 25, 26
Fejes T.L., 25–26
Fellner, F., 151
Fermi, E., 33, 114
Fermi, L., 172
Fizikai Szemle (Hungarian Physical Review), 103
Fodor, L., 120
Folk art, 21
Fomenko, A.T., 63
Food and Drug Administration (FDA, USA), 6
Francis Crick Institute, 119
Franklin, R., 9
French Academy of Sciences, 129, 130
Fritz Haber Institute, 133
Frost, R., 135
Fuller, R.B., 58, 59
Furchgott, R.F., 34
Furka, A., 27–28

G
Gabor, D., 135
Gadó, P., 29–32
Galilei, G., 75
Gamow, G., 9, 11
Gardner, M., 73, 74
Garwin, R.L., 33–36
Gell-Mann, M., 118
Geodesic Dome, 58
George A. Olah Lecture, 23, 121
Giger, H., 64
Gillespie, R.J., 37–39
Ginzburg, V.L., 141
Godlee Lecture, 9, 10
Göncz, Á., 45–49, 164
Goodfriend, A., 41–43
Gorbachev, M.S., 141
Great Minds (Hargittai, Hargittai and Hargittai), 147
Guba, F., 149

H
Halápy, J., 66, 67
Hamlet (Shakespeare), 121
Hanson, H.P., 2, 133
Harvard University, 70, 129, 130, 156
Hassel, O., 2, 27
Heisenberg, W., 155
Heller, Á., 48
Hermitage, 20

Hernádi, J., 51–52
Herschbach, D., 133
Hershko, A., 46, 54–56
Hershko, J., 53
Higgs, P., 118
History of the Academy of Sciences, 28, 46, 150
Hitler, A., 110, 114, 134
Hodgkin, D., 149
Hoffmann, R., 49
Holocaust, 26, 46, 48, 54, 70, 94, 98, 150, 151
Horthy era, 16, 94, 150
Horthy, N., 42, 70, 105, 134, 151
Horthy regime, 16, 28, 29, 36, 46, 103, 114, 134, 135, 150, 151, 173
Human relations, xiv
Human rights, 139, 141
Human values, 14
Hungarian Consulate General in New York, 43
Hungarian Nobel laureates, 23, 35, 69, 104, 149
Hungarian–Norwegian interactions, 3
Huzella, T., 7
Hydrogen bomb, 33, 140, 141, 154, 156, 157

I
Illustrious Immigrants (L. Fermi), 172
Illyés, G., 16, 17
Immigration, 30, 104
Imperial College Press, 34, 38, 145, 147
Innovations, 8, 51, 52, 90, 102, 108, 140
Institute of Crystallography (Moscow), 126
Internal emigration, 101
International Herald Tribune, 128
International Union of Crystallography, 99
Interviews, 2, 31, 32, 38, 110, 130, 143, 145, 146, 164
Introduction to Stereochemistry (Mislow), 109

J
Jacob, F., 11
Jewish Hospital, 24
Jewish resistance, 118
Jewish school, 54
John Paul II, 10
Johnson, L.B., 173
Journal of Chemical Education, 20, 21
Judging Edward Teller (Hargittai), 154

K
Kádár regime, 48
Kahn, L., 58–60, 74, 85
Kaldor, N., 135
Kant, I., 111
Karolinska Institute, 7, 55, 69, 83
Karrer, P., 35
Kazár, J., 168
Kemeny, J., 103
Kepes, A., 33
Kepes, G., 61–64
Kepler, J., 75, 102
Kertész, I., 71
Kerti, K., 65–68
KGB, 141, 144
Khrushchev, N.S., 80, 141
Kindertransport (Kent), 151
Kitaigorodsky, A., 47, 125

Klein, E., 69, 71
Klein, G., 24, 36, 49, 69–72
Kodak (company), 30
Koenig, F., 67, 68
Koestler, A., 73–75, 154
Komoróczy, G., 175
Kornfeld, K., 93
König, D., 93
Konrád, G., 55, 56
Kövesdy, Pál (Paul Kovesdy), 67

L
Laboratory of Molecular Biology (Cambridge, UK), 36
Lagers (camps), 26, 46, 94
Landau, L.D., 30, 80
Lantos, F., 19, 20, 77–81, 138
Larson, C., 145
Larson, J., 145
Laurent, T.C., 7, 23, 83–86
Lauterbur, P., 89–91
Lax, H., 93
Lax, L., 33
Lax, P., 33, 93–95
Lázár, D., 25, 26
Lehn, J.-M., 129, 131
Lendvai, E., 128
Lengyel, Piroska, 98
Lengyel, B., 98
Lengyel, G., 21
Lengyel, O., 97
Lengyel, S., 97–98
Lenin, V.I., 15
Levine, D., 102
Levine, J.L., 155
Levi, P., 110
Lichtenstein, R., 78, 79
Ligeti, G., 71
Lomonosov University, 1, 15, 63, 80, 107, 140, 159, 160, 172
Lord Kelvin, 111
Los Alamos, 33, 41, 94, 95
Louis Pasteur University, 129
Lövei, M., 169
Lowrey, A., 78
Lukacs, G., 48
Lysenko, T.D, 80

M
Mackay, A.L., 77, 85, 99–102
Magnetic resonance imaging (MRI), 36, 89–91
Mamedov, K., 126, 127
Manchester University, 135
Manhattan Project, 114, 118, 155
Mansfield, P., 89–91
Martians, 94, 114, 115, 126, 133, 134, 156
Martians of Science (The Martians of Science), 105, 114
Marx, G.(György), 103–105
Masaryk University, 150
Massachusetts Institute of Technology (MIT), 61
Mathematical Intelligencer, The, 137, 138
Matisse, H., 20
Matrix Biology Institute, 7, 8
Mauthner, N., 151
Max-Planck-Gesellschaft, 133
Mechanism of reactions, 121, 133, 135, 143

Medical Research Council (British), 35, 36
Memorials, 3, 16, 26, 47, 48, 68, 80, 104, 108, 129, 130, 143, 150, 151, 155
Mentors, 25, 30, 47, 54, 55, 84, 135, 155, 160, 168
Meselson, M., 156, 164
Metropolitan Museum (New York), 67
Mezei, A., 56
Mez-Starck, B., 107–108
Mislow, K., 109–111
Models/modeling, 9, 10, 20, 21, 30, 37–39, 58, 61, 75, 93, 94, 102, 125, 133
Moiré patterns, 64
Moncada, S., 35
Munger, C.T., 114–115
Musée d'Orsay, 20
Museum of Modern Art (New York), 20
Mussolini, B., 110, 150

N
Nakanishi, K., 31, 32
National Institute of Standards and Technology (NIST), 99
National Science Foundation (NSF), 111
National socialism (Nazism), 134
Nature, 90, 138
Naval Research Laboratory (NRL), 78
Ne'eman, Y., 117–118
Neizvestny, E., 80
Nesmeyanov, A.N., 15
New School (New York), 48
New York Hungarian Scientific Society, 43
New York Scientific (Hargittai and Hargittai), 43, 48
New York University (NYU), 35, 93, 94, 111
Newton, I., 75
Nitrogen oxide (NO), 35
Nixon Administration, 173
Nobel Committee of Chemistry, 24, 84, 138
Nobel Forum, 94
Nobel Foundation, 84, 122
Nobel Peace Prize, 140
Nobel Prize Committee, 140, 156
Norrish, R.G.W., 101
Norwegian Academy of Science and Letters, 3
Norwegian Research Council, 2
Novák, B., 120
Novodeviche Cemetery, 80, 144
Numerus clausus, 23, 70, 94, 134, 150
Nurse, P., 119–120
Nye, M.J., 133
Nyholm, R.S., 38

O
Ochoa, S., 35
Odd Hassel Lecture, 2
Olaf, V., 3
Olah, G.A., 38, 83, 121–123, 138
Olah, J. (née Lengyel), 122, 155
Olah, P., 122
One sentence about tyranny (Illyés), 16
On Rasor's Edge (Maugham), 70
On Tyranny (Snyder), 16, 17
Oppenheimer hearing, 154
Oppenheimer, J.R., 154
Orosz, I., 3, 125–128
Ortvay, R., 173

Orwell, G., 73
Oslo University, 1, 2, 140
Ourisson, G., 129–131
Our Lives (Hargittai), 45–49, 164, 165
Oxford University, 36, 86, 101, 114, 147
Oxford University Press, 86, 114, 147

P
Paintings/graphics, 27, 63, 66, 67, 78, 79, 125, 128
Pais, A., 134
Pasteur, L., 111, 129
Patents, 6, 7, 90
Pauling, L., 100, 110, 138
Peace through Chemistry (Lichtenstein), 78, 79
Penrose, R., 101, 102
Penzias, A., 151
Pericles, 14
Persecution of Jews, 46, 110
Personal Knowledge (Polanyi), 133
Péter, R., 93
Pharmacia (company), 6
Philosophy, 41, 74, 133–135
Physical Institute (Moscow), 140, 141
Polanyi, J.C., 133–135
Polanyi, M., 2, 30, 133–135, 172
Pólya, G., 94, 135
Popularizing science, 38, 73, 103, 114, 143
Premature discoveries, 135
Prigogine, I., 2
Princeton University, 109, 110
Prostaglandins, 6

Q
Quasicrystals, 77, 85, 102, 137

R
Racial discrimination, 111
Racism, 164
Radda, G., 36
Radio Budapest, 143
Radnóti, M., 129, 130, 172
Radnóti, S., 71
Radulescu, G., 123, 138, 144
Rákosi dictatorship, 5
Regan, L., 114, 115
Religion, 9, 10, 98, 140
Rényi Institute of Mathematics, 26
Research Laboratory of Structural Chemistry, 1, 51, 97, 98
Richter, G., 151
Righteous Among the Nations, 98
Rockefeller University, 119
Rodin, E.Y., 62
Roosevelt, F.D., 78
Rose, I., 54
Royal Society (London), 102, 119, 134
Royal Swedish Academy of Sciences, 23, 24, 84–86
Russian Academy of Sciences, 164

S
Sacks, O., 48
Sakharov, A.D., 41, 139–141
Sarkadi, K., 108

Science history, 28, 32, 35, 89, 164
Science of structures, 102
Scientific American, 58, 73
Seaborg, G.T., 145
Semenov, N., 143–145
Shechtman, D., 77, 85, 99–102
Shelter Publications, 58
Shustorovich, E., 29, 30
Simonffy, G., 143–147
Slave labor, 5, 15, 24, 26, 46, 54, 66, 67, 129
Snelson, K., 59, 60
Snyder, T., 16, 17
Somorjai, G., 36
Soros Stipend, 161
Soviet Academy of Sciences (Russian Academy of Sciences), 140
Stalin, J.V., 15, 16, 24, 47, 73, 80, 114, 141, 154, 157
Stalin's terror, 16, 73, 80
Stanford University, 94
Starck, H.C., 108
Star Wars concept, 34
State University of New York (SUNY), 34
Steiner, L., 151
Steinhardt, P., 102
Stern, M., 46
Stockholm University, 23, 24, 84
Strasshof, 54
Structural chemistry, 1, 51, 89, 97, 98, 102, 123, 126
Superacids, 38
Symmetries
    Escher, 126
    Fejes Tóth, 25, 26, 54
    Fomenko, 63
    Gardner, 73, 74
    Kepes, 61, 62
    Koenig, 68
    Lantos, 77, 78, 138
    Laurent, 7, 23, 83–85
    Lendvai, 128
    Lengyel, Györgyi, 21
    Mackay, 77, 85, 99, 100
    Mamedov, 126
    Mislow, 109–111
    Ne'eman, 117, 118
    Neizvestny, 80, 81
    Olah, 121–123, 138
    Orosz, 125, 126
    Shechtman, 77, 85, 99, 101, 102
    Vasarely, 127
    Wigner, 2, 29, 30, 103, 114, 126, 133, 134, 172, 173
    Witschi, 64
Symmetry (journal), 127
Symmetry: A Unifying Concept (Hargittai and Hargittai), 57
Symmetry through the Eyes of a Chemist (Hargittai and Hargittai), 21
Symmetry 2000 (Hargittai and Laurent), 85
Symmetry: Unifying Human Understanding (Hargittai), 61
Synergetics (Fuller), 59
Szavaink születése (Birth of Our Words, Dalos), 16
Széchenyi I.G., 150
Szegő, G., 94
Szemerédi, E., 33
Szent-Györgyi, Albert, 149–151
Szent-Györgyi, Andrew, 149
Szilard, L., 11, 103, 114, 126, 156, 165
Sztehlo, G., 122

T
Tamm, I., 140, 141
Technion (Israel Institute of Technology), 117
Telegdi, V., 34
Teleki, P., 70
Teller, E., 33, 34, 73, 93, 94, 105, 114, 133, 153–157
Tensegrity, 59, 60
Természet Világa (World of Nature), 38
Terror Háza (House of Terror), 94
The Double Helix (Watson), 164
The Loyal Subject (Der Untertan), 70
The Martians of Science (Hargittai), 105, 114
The Road to Stockholm (Hargittai), 10, 71, 86, 114
The Sleepwalkers (Koestler), 75
The Sphere (Koenig), 67, 68
The Union Jack (Kertész), 71
The Voice of the Dolphins (Szilárd), 126
The Watershed (Koestler), 75
Thucydides, 14
Totalitarian regimes, 29, 73
Truncated icosahedron, 58
Tulane University, 110

U
Uitz, B., 16
Ulam, S., 33
Unity (Mamedov), 126, 127
University of
    Birmingham, 120
    California, 35
    Connecticut, 61, 67, 75, 110, 138
    East Anglia, 120
    Florida, 2
    Freiburg, 107, 108
    Illinois, 90
    Kentucky, 5
    London, 5, 9, 10, 101
    Nottingham, 89, 90
    Southern California, 23, 83, 121, 123
    Szeged, 28, 173
    Texas, 2, 37, 115, 133, 171, 172
    Toronto, 123
    Ulm, 31, 108
    Uppsala, 84
    Zurich, 26, 35
Upptäck symmetri! (Hargittai and Hargittai), 85

V
Valence shell electron pair repulsion (VSEPR), 37, 38
Vane, J., 119
Varga, J., 16
Vasarely, V., 3, 128
Vienna, 35, 47, 48, 54, 66, 151
Vienna University, 35
Vilkov, L.V., 159–161
Vision + Value (Kepes), 61
Vitamin C, 149
von Kármán, T., 93, 114, 133
von Neumann, J., 26, 29, 94, 114, 126, 133, 149, 172, 173
Vörösmarty, M., 172
Votava, G., 48
Vrba, R., 70

W
Wallenberg, R., 23, 24, 36
Watson, J.D., 9, 45, 164–165
Weinberg, S., 118
Weissmann, C., 35
Wenner-Gren Foundation, 23, 84, 85
Wetzler, A., 70
Wheeler, J., 134
Wiegandt, R., 167–169
Wigner, E.P., 2, 29, 30, 103, 114, 133, 134, 172–173
Wilhelm, A., 26
Wilhelm, J., 26
Williams, W.C., 74
Wisdom of the Martians (Hargittai és Hargittai), 105
Witschi, W., 64
Wohl, E., 23

Wohl, H., 24
Women scientists (M. Hargittai), 147
World Trade Center (New York), 68

X
Xalatan, 5–8

Y
Yad Vashem, 98

Z
Zemplén, G., 122
Zsidósors Magyarországon (Jewish Fate in Hungary, Lévai), 94

Printed in the United States
by Baker & Taylor Publisher Services